A LIFE OF
MAGIC CHEMISTRY

A LIFE OF MAGIC CHEMISTRY

Autobiographical Reflections of a Nobel Prize Winner

George A. Olah

WILEY-INTERSCIENCE

A JOHN WILEY & SONS, INC., PUBLICATION

New York • Chichester • Weinheim • Brisbane • Singapore • Toronto

Copyright © 2001 by Wiley-Interscience. All rights reserved.

Published simultaneously in Canada.

For ordering and customer service call 1-800-CALL-WILEY.

Library of Congress Cataloging-in-Publication Data:

Olah, George A. (George Andrew), 1927–
    A life of magic chemistry : autobiographical reflections of a nobel prize winner /
George A. Olah.
       p. cm.
    Includes bibliographical references and index.
    ISBN 0-471-15743-0 (cloth : alk. paper)
       1. Olah, George A. (George Andrew), 1927–. 2. Chemists—United States—Biography.
    I. Title.
    QD22.043 A3 2000
    540'.92—dc21
    [B]   00-043638

Printed in the United States of America.

10 9 8 7 6 5 4 3 2 1

To Judy,
who made it all possible

My grandchildren, Peter and Kaitlyn (July 1999).

Contents

Preface

My wife Judy, my children, and my friends urged me for some time to write about my life and the fascinating period of science I was lucky to be part of. For years I resisted, mainly because I was still fully occupied with research, teaching, and various other commitments. I also felt it was not yet time to look back instead of ahead. However, I slowly began to realize that, because none of us knows how much time is still left, it might be ill advised to say "it is not yet the right time." I therefore started to collect material and to organize my thoughts for a book.

It soon became clear that this project would be very different from any writing I had done before. I recognized that my goal was not only to give autobiographical recollections of my life and my career in chemistry but also to express some of my more general thoughts. These touch on varied topics, including the broader meaning of science in the quest for understanding and knowledge as well as their limitations. Science as a human endeavor means the search for knowledge about the physical world. Inevitably, however, this leads to such fundamental questions of how it all started and developed: Was there a beginning? Was our being planned by a higher intelligence? We struggle with these and related questions while trying to balance what we know through science and what we must admit is beyond us. My thoughts are those of a scientist who always tried to maintain his early interest in the classics, history, philosophy, and the arts. In recent years I have particularly tried to fill in some of the gaps; a life actively pursuing science inevitably imposes constraints on the time that one can spend reading and studying outside one's own field of specialization. Of course, I realize only too well my limitations and the lack of depth in my background in some of these areas. Therefore, I have tried not to overreach, and I will limit my thoughts to my own under-standing and views, however imperfect they may be.

This book is mainly about my life in search of new chemistry. Because some of my work centered around the discovery of extremely strong "superacids," which are sometimes also called "magic acids," I chose the title *A Life of*

Magic Chemistry. It also reflects in a more general way the exciting and sometimes indeed even "magic" nature of chemistry, which with its extremely broad scope cuts through many of the sciences, truly being a central science.

It was a long journey that led me from Budapest through Cleveland to Los Angeles with a side trip to Stockholm. Sometimes I still wonder how life unfolds in ways we could not have planned or foreseen.

I thank my publisher for the patience and understanding shown for my delays in writing the book. My editors Darla Henderson, Amie Jackowski Tibble, and Camille Pecoul Carter helped greatly to make the book a reality. My wife, sons, and friends helped to improve the manuscript and commented on its many shortcomings. My particular thanks go to Reiko Choy, my longtime secretary, who, before her retirement, miraculously produced a proper manuscript from my messy handwritten drafts and thus allowed the book to be completed. I similarly thank Jessie May, who took over and carried through with great efficiency and enthusiasm needed revisions and corrections.

George A. Olah
Los Angeles, October 2000

· 1 ·

Introduction

If we look back on the history of human efforts to understand our world and the universe, these look like lofty goals that, I believe, mankind will never fully achieve. In earlier times, things were more simple. The great Greek thinkers and those who followed in their footsteps were able to combine the knowledge available of the physical world with their thoughts of the "spiritual world" and thus develop their overall philosophy. This changed with the expansion of scientific inquiry and quest for knowledge in the seventeenth century. By the twentieth century, few philosophers, except those who also had some background in the sciences, could claim sufficient knowledge of the physical world to even attempt serious consideration of its meaning. This opened the claim to center stage to some scientists, particularly physicists, suggesting that only science can attempt to give answers to such fundamental questions as the origin and meaning of the universe, life, our being as intelligent species and the understanding of the universal laws governing the physical and biological world. In reality, however, humankind with all its striving for such knowledge probably will never reach full understanding. For me this is readily acceptable. It seems only honest to admit our limitations because of which human knowledge can reach only a certain point. Our knowledge will continue to expand, but it hardly can be expected to give answers to many of the fundamental questions of mankind. Nuclear scientists developed insights in the ways in which the atoms of the elements were formed after the initial "big bang," but chemists are concerned with the assembly into molecules (compounds, materials) and their transformations. They can avoid the question of whether all this was planned and

1

created with a predetermined goal. I will, however, briefly reflect on my own views and thoughts. They reflect my struggle and inevitable compromises, leading to what I consider—at least for me—an acceptable overall realization that we, in all probability, never can expect a full understanding.

I was lucky to be able to work during and contribute to one of the most exciting periods of science, that of the second half of the twentieth century. I was also fortunate that I was mostly able to pursue my interests in chemistry, following my own way and crossing conventional boundaries. Frequently, I left behind what Thomas Kuhn called safe, "normal science" in pursuit of more exciting, elusive new vistas. How many people can say that they had a fulfilling, happy life doing what they love to do and were even paid for it? Thus, when people ask me whether I still work, my answer is that I do, but chemistry was never really work for me. It was and still is my passion, my life. I do not have many other interests outside chemistry, except for my family and my continuous learning about a wide range of topics through reading. Thus the long hours I still spend on science come naturally to me and are very enjoyable. If, one day, the joy and satisfaction that chemistry gives me should cease or my capabilities decline so that I can make no further meaningful contributions (including helping my younger colleagues in their own development and efforts), I will walk away from it without hesitation.

In recent years, I have also grown interested in attempting to link the results of my basic research with practical uses done in environmentally friendly ways. This means finding new ways of producing hydrocarbon fuels and derived materials and chemicals that at the same time also safeguard our fragile environment. Pinpointing environmental and health hazards and then regulating or, if possible, eliminating them is only one part of the question. It is through finding new solutions and answers to the problems that we can work for a better future. In this regard chemistry can offer much. I find it extremely rewarding that my colleagues and I can increasingly contribute to these goals in our field. This also shows that there is no dichotomy between gaining new knowledge through basic research and finding practical uses for it. It is a most rewarding aspect of chemistry that in many ways it can not only contribute to a better understanding of the physical and biological world but also supplement nature by allowing man to produce through his own efforts essential products and materials to allow future generations a better life while also protecting our environment.

· 2 ·

Perspectives on Science

I have spent my life in science pursuing the magic of chemistry. In attempting to give some perspectives and thoughts on science, it is first necessary to define what science really is. As with other frequently used (or misused) terms (such as "God" or "democracy") that have widely differing meanings to different people at different times and places, "science" does not seem to be readily and uniformly defined. Science, derived from the Latin "scientia," originally meant general knowledge both of the physical and spiritual world. Through the ages, however, the meaning of science narrowed to the description and understanding (knowledge) of nature (i.e., the physical world). Science is thus a major intellectual activity of man, a search for knowledge of the physical world, the laws governing it, and its meaning. It also touches on fundamental, ageless questions as to our existence, origin, purpose, and intelligence and, through these, the limits of how far our understanding can reach. In many ways scientists' intellectual efforts to express their thoughts and quest for general knowledge and understanding are similar to other intellectual efforts in areas such as the humanities and arts, although they are expressed in different ways.

In discussing science we also need to define its scope, as well as the methods and views (concepts) involved in its pursuit. It is also useful to think about what science is *not*, although this can sometimes become controversial. Significant and important studies such as those concerned with the fields of sociology, politics, or economics increasingly use methods that previously were associated only with the physical and biological sciences or mathematics. However, I believe these

4

are not in a strict sense "hard sciences." The name "science" these days is also frequently hyphenated to varied other fields (from animal-science to culinary science to exercise science, etc.). Such studies indeed may use some of the methods of science, but they hardly fall under the scope of science. There is a Dutch proverb that says "Everything has its science, with the exception of catching fleas: This is an art." It may overstate the point, but sometimes to make a point it is necessary to overstate it.

When we talk about *knowledge* of the physical world, we generally refer to facts derived from systematic observation, study, and experimentation as well as the concepts and theories based on these facts. This is contrasted with *belief* (faith, intuition) in the spiritual or supernatural.

Scientists use methods in their pursuit of knowledge that frequently are referred to collectively as the "scientific method," originally credited to Francis Bacon dating from the end of the sixteenth century. Bacon believed that the facts in any given field can be collected according to accepted and prearranged plans and then passed through a logical intellectual process from which the correct judgments will emerge. Because phenomena (facts) were so numerous even then, he suggested that they must be chosen (selected), which is a subjective act of judgment. This process is hardly compatible with what we now associate with the scientific method.

This also brings up the essential relationship of science and its historical perspective. We can never talk about science without putting it into a time frame. August Comte wrote, "L'histoire de la science c'est la science meme"—"The history of science is really science itself." When we look back in time early scientists (savants) long believed that the earth was the center of the universe and that it was flat. They even warned that approaching its edges would put one at risk of falling off. However strange this may be for us today, they were interpreting the limited knowledge they had at the time. We may pride ourselves on what we consider our advanced knowledge as we enter the twenty-first century, but I am sure future generations will look back at us and say how ignorant and naïve we were. As Einstein said, "One thing I have learned in a long life is that all of our science, measured against reality, is primitive and child-like and yet it is the most precious thing we have." I hope that it will also be remembered that we tried our best. Scientific knowledge by its nature continuously changes and expands. Only through its historical time frame can science be put into its proper perspective. It is thus regrettable that the history of science is not taught in many of our universities and colleges. This probably is also due to the fact that the interactions between scientists and historians (philosophers), and the mutual understanding of the significance of their fields, are frequently far from satisfactory.

The days are long gone when friends of Lavoisier, one of the greatest scientists of all time, during the terror of the French revolution, were pleading for his life before the revolutionary tribunal, which, however, ruled that "la revolution n'a pas besoin de la science" (the revolution does not need science). He went to the guillotine the same day. Since that time it has become clear that the world needs science for a better future. Science does not know national, racial, or religious distinctions. There is no separate American, European, Chinese, or Indian science; science is truly international. Although scientific results, like anything else, can also be misused (the use of atomic energy is still frequently condemned because its development was closely related to that of the atom bomb), we cannot be shortsighted and must look at the broader benefits of science.

The scientific method, as mentioned, involves observation and experimentation (research) to discover or establish facts. These are followed by deduction or hypothesis, establishing theories or principles. This sequence, however, may be reversed. The noted twentieth-century philosopher Karl Popper, who also dealt with science, expressed the view that the scientist's work starts not with collection of data (observation) but with selection of a suitable problem (theory). In fact, both of these paths can be involved. Significant and sometimes accidental observations can be made without any preconceived idea of a problem or theory and vice versa. The scientist, however, must have a well-prepared, open mind to be able to recognize the significance of such observations and must be able to follow them through. Science always demands rigorous standards of procedure, reproducibility, and open discussion that set reason over irrational belief.

Research is frequently considered to be either basic (to build up fundamental knowledge) or applied (to solve specific practical goals). I myself have never believed in a real dividing line. Whenever I made some new basic findings in chemistry I could never resist also exploring whether they might have a practical use. The results of scientific research can subsequently be developed into technology (research and development). It is necessary to differentiate science from technology, because they are frequently lumped together without clearly defining their differences. To recapitulate: *Science* is the search for knowledge; *technology* is the application of scientific knowledge to provide for the needs of society (in a practical as well as economically feasible way).

"In the pursuit of research or observation many would see what others have seen before, but it is the well-prepared one who [according to Albert Szent-Györgyi, Nobel Prize in medicine 1937] may think what nobody else has thought before" and achieve a discovery or breakthrough. Mark Twain once wrote that "the greatest of all inventors is chance." Chance, however, will favor only those who are capable of recognizing the significance of an unexpected invention and explore it further.

Thomas Kuhn, the science philosopher, in his *Structure of Scientific Revolutions*, called "normal science" research that is based upon established and accepted concepts (paradigms) that are acknowledged as providing the foundation for the future. This is the overwhelming part of scientific research. It is also considered "safe" to pursue because it is rarely controversial. Following Yogi Berra's advice, it allows the scientists "not to make the wrong mistakes." Consequently, it is usually also well supported and peer approved. Some scientists, however, dare to point out occasionally unexpected and unexplained new findings or observed anomalies. These always are "high risk" and controversial and frequently turn out to be flukes. But on occasion they can lead to new fundamental scientific discoveries and breakthroughs that advance science to new levels (paradigm changes). Kuhn called this "revolutionary science," which develops when groundbreaking discoveries cannot be accommodated by existing paradigms.

Science develops ever more rigorous standards of procedure and evaluation for setting reason aside from irrational belief. However, with passing time and accumulated knowledge many concepts turn out to be incorrect or to need reevaluation. An example mentioned is the question of earth as the center of our universe. Others range from Euclidean geometry to the nature of the atom.

Euclid's fifth axiom is that through every point it is possible to draw a line parallel to another given line. This eventually turned out to be incorrect when it was realized that space is curved by gravity. The resulting non-Euclidian geometry became of great use and was applied by Einstein in his general theory of relativity. Kant believed that some concepts are a priori and we are born with them: all thought would be impossible without them. One of his examples was our intuitive understanding of three-dimensional space based on Euclidean geometry. However, Einstein's space-time fourth dimension superseded Euclidean geometry.

One of the characteristics of intelligent life that developed on our planet is man's unending quest for knowledge. (I am using "man" as a synonym for "humans" without gender differentiation.) When our early ancestors gazed upon the sun and the stars, they were fascinated with these mysterious celestial bodies and their movement. Ever since, man has strived to understand the movement of heavenly bodies. But it was only such pioneers as Copernicus, Kepler, and Galileo who established the concepts of celestial mechanics, which eventually led to Newton's theory of gravitation. Physics thus emerged as a firm science in the seventeenth century.

Contrasted with the mind-boggling, enormous scale of the cosmos, our understanding of the atomic nature of matter and the complex world of infinitesimally small subatomic particles and the forces within the atom presents another example for our continuously evolving and therefore changing knowledge. Starting with the early Greek atomists it was believed that the universe was made up of atoms, the further undividable elemental matter. The past century saw, however, an explosive growth in our knowledge of subatomic particles. The recognition of the electron, proton, and neutron was followed by the discovery of quarks and other subatomic particles.

In the nineteenth century, scientists showed that many substances, such as oxygen and carbon, had a smallest recognizable constituent that, following the Greek tradition, they also called atoms. The name stuck, although it subsequently became evident that the atom is not indivisible. By 1930, the work of J. J. Thomson, Ernest Rutherford, Niels Bohr, James Chadwick, and others established a solar system-like atomic model consisting of a nucleus containing protons and neutrons and surrounded by orbiting electrons. In the late 1960s it was shown that protons and neutrons themselves consist of even smaller particles called quarks. Additional particles in the universe are the electron-neutrino (identical to the electron but 200 times heavier), the muon, and an even heavier analog of the electron called tau. Furthermore, each of these particles has an antiparticle identical in mass but of opposite charge. The antiparticle of the electron is the positron (with identical mass but with a charge of $+1$ instead of -1). Matter and antimatter, when in contact, substantially (but not necessarily completely) annihilate each other. This is the reason why there is extremely little antimatter around and it is so difficult to find.

Besides particles, the forces of nature play also a key role. In the past century four fundamental forces were recognized: the gravitational, electromagnetic, weak, and strong forces. Of these the weak and strong forces are less familiar, because they are nuclear forces and their strength rapidly diminishes over all but subatomic scales.

During Einstein's time the weak and strong forces were not yet known. However, gravity and electromagnetism were recognized as distinct forces. Einstein attempted to show that they are really manifestations of a single underlying principle, but his search for the so-called unified field theory failed. So did all efforts to combine the two major pillars of modern physics, quantum mechanics, and general relativity. As presently formulated both cannot be right because they are mutually incompatible. Attempts are being made to find a unified theory for everything, to prove that there is one set of laws for the very large things and the smallest alike, including all forces and particles. Although physicists long believed that the minuscule electrons, quarks, etc. are the smallest particles of matter, the recently pursued string theory suggests that there is an even deeper structure, that each elementary particle is a particular node of vibration of a minute oscillating string. The image replacing Euclid's perfect geometric points is that of harmoniously thrumming strings (somewhat like Pythagoras' music of the spheres). These infinitesimal loops or strings are suggested as writhing in a hyperspace of 11 dimensions. Of these only four dimensions are easily comprehended by us, the three dimensions of space and Einstein's space-time. The seven additional dimensions of the superstring theory (or as it is sometimes called, the *theory of everything*) are "rolled up" or "compacted" into an infinitesimally small format but are still not dimensionless points. The principle that everything at its most microscopic level consists of a combination of vibrating strands of strings is the essence of the unified theory of all elemental particles and their interactions and thus all the forces of nature.

The complex mathematical basis of the string theory is far beyond the understanding of most of us, and certainly beyond my understanding. However impressive and elegant the mathematical *tour de force* may be that one day could produce an "equation for everything" containing 11 dimensions, it is not clear what its real meaning will be. This is a difficult question to ponder. The tiny domain that superstrings inhabit can be visualized by comparing the size of a proton to the size of the solar system. The entire solar system is 1 light day around, but to probe the reality of the tiny realm of superstrings would require a particle accelerator 100 light years across (the size of our solar system). As long as the superstring theory or any of its predictions that may emerge cannot be experimentally tested (or disproved), it will remain only a mathematical theory. However, the progress of science may one day result in ingenious new insights that can overcome what we presently perceive as insurmountable barriers.

John von Neuman, one of the greatest mathematicians of the twentieth century, believed that the sciences, in essence, do not try to explain, they hardly even try to interpret; they mainly make models. By a model he meant a mathematical construct that, with the addition of certain verbal interpretations, describes observed phenomena. The justification of such a mathematical construct is solely and precisely that it is expected to work. Stephen Hawking also believes that physical theories are just mathematical models we construct and that it is meaningless to ask whether they correspond to reality, just as it is to ask whether they predict observations.

For a long time, views and concepts (theories) of science were based on facts verified by experiments or observations. A contrary view was raised by the philosopher Karl Popper, according to whom the essential feature of science is that its concepts and theories are not verifiable, only falsifiable. When a concept or theory is contradicted by new observations with which it is incompatible, then it must be discarded. Popper's views were subsequently questioned (Kuhn, Feyerabend) on the basis that falsification itself is subjective, because we do not really know a priori what is true or false. Nonetheless, many still consider "scientific proof," i.e., verification, essential. Gell-Mann (Nobel Prize in physics 1969), for example, writes in his book, *The Quark and the Jaguar*, "sometimes the delay in confirming or disproving a theory is so long that its proponent dies before the fate of his or her idea is known. Those of us working in fundamental physics during the last few decades have been fortunate in seeing our theoretical ideas tested during our life. The thrill of knowing that one's prediction has been actually verified and that the underlying new scheme is basically correct may be difficult to convey but is overwhelming." Gell-Mann also wrote "It has often been said that theories, even if contradicted by new evidence, die only when their proponents die." This certainly may be the case when forceful personalities strongly defend their favorite brainchildren. Argumentum ad hominem, however, does not survive for long in science, and if a theory is superseded just because its proponent is not around any more to fend off the others questioning it, it surely sooner or later will be "falsified."

Gell-Mann seems to believe that scientific theories are verifiable and can be proven (confirmed) even in one's own lifetime and thus proven to be true. This is, however, not necessarily the general case as, for example, his own quarks may turn out not to be the ultimate elementary particles. Recent, tentative experimental observations as well as theory seem to cast doubt on the idea that quarks are indeed the smallest fundamental, indivisible particles of the atoms. They themselves are probably made up of even smaller entities of yet-unknown nature. As discussed, the superstring theory suggests that all matter, including quarks, is composed of vibrating strings. Whereas quarks may stay on for the time being as the fundamental particles, future work probably will bring further understanding of atomic physics with even more diverse particles and forces being recognized.

"Theory" means the best possible explanation of observations, experimental facts, or concepts (hypotheses) as we know them or conceive them at the time. If new observations (facts or concepts) emerge with which the theory cannot be in accord, then we need to discard or modify the theory. Theories thus cannot be absolutely verified (proven) or even falsified (disproved). This should not imply, however, that a discarded theory was necessarily incorrect at the time it was proposed or represented any intent to deliberately mislead or misrepresent. As I have emphasized, science can never be considered without relating it to its historical time frame. There is continuing progress and change in our scientific concepts as new knowledge becomes available. Verification or proof of a theory in the present time thus may be only of temporary significance. Theories can be always superseded by new observations (facts) or concepts. This is the ongoing challenge of science.

The widely invoked concept of "chaos" based on chaotic phenomena is, by our present understanding, unpredictable. According to Ilya Prigogine (Nobel Prize in chemistry 1977), we have reached the end of certitude in science, which in the future will be increasingly speculative and probabilistic (i.e., ironic). Others, however, feel that eventually a deeper new understanding of some yet-unknown law governing chaotic phenomena will be found. The question is, when are we really reaching the limits of real understanding or knowledge? Are vibrating infinitesimally small strings indeed the basis of all matter and forces, allowing a "theory of everything" eventually to be found? Is our universe just one of innumerable multiverses? Is evolution a conscious, predetermined process making the emergence of intelligent beings inevitable or just a consequence of nature? And, ultimately, why is there anything, did it all start and will eventually come to an end, or was it always and always will be? Creation means a beginning, but it is possible to think in terms of a continuum without beginning or end. Science in all probability cannot and will never be able to answer these questions. To me, it is only honest to admit that we just don't know.

However, one can go too far in delineating science, as did Thomas Kuhn, who contended that all science reflects not the truth about nature but merely the scientists' prevailing opinion, which is always subject to change. Science has, however, established many fundamental observations and facts of our physical world. For example, atoms exist in a variety corresponding to the elements, as do DNA, bacteria, stars and galaxies, gravity and electromagnetism, natural selection and evolution. Science is our quest for understanding of the physical world, and we should keep this in proper perspective while admitting to the limits of where our human understanding can reach.

A fundamental question in our quest for knowledge and understanding always will be whether there is a higher intelligence beyond our grasp. Many call this "God," but that name invokes very different meanings to different people. It seems that in many ways man created God in his own image or at least depicted him accordingly. Scientists in general find it difficult to believe in something they cannot comprehend or understand. I myself have found it increasingly difficult over the years to believe in supernaturals as proclaimed by many organized religions and their dogmas and regulations. Monotheism is accepted in Judaism, Christianity, and Islam, but there are also other religions such as Buddhism, Confucianism, Hinduism, and Taoism, among others.

The Scriptures of the Bible, as well as the Talmud, and the Book of Mormon, are all valuable teachings and worthy historical documents, but much in them can hardly be taken verbatim. For example, creation according to the Book of Genesis has a limited time line that cannot readily reconciled with scientific knowledge of our physical and biological world. At the same time, science itself cannot give an answer to how it all began ex nihilo (if it started at all). The "big bang" that happened 12–15 billion years ago only explains how our expanding universe probably started from an immensely dense initial state, not how this came about. We seem to increasingly comprehend how subsequent inflation and continued expansion are governed by physical laws. But there may be innumerable other universes, too, which are not necessarily governed by the same physical laws as ours.

All recognized religions are by necessity quite contemporary. What is a few thousand years compared to what we know of how long life and even humankind have been around on earth? Disregarding the enormous time discrepancy of the biblical act of creation with existing scientific evidence of life on earth, an omnipotent god with a definite act of creation simplifies many questions for true believers. There are also many other questions, such as those of our consciousness and free will, whether there was indeed a beginning, whether there is a reason or goal of our being, and was it planned, to which science itself cannot give answers. Today, I consider myself, in Thomas Huxley's terms, an agnostic. I don't know whether there is a God or creator, or whatever we may call a higher intelligence or being. I don't know whether there is an ultimate reason for our being or whether there is anything beyond material phenomena. I may doubt these things as a scientist, as we cannot prove them scientifically, but at the same time we also cannot falsify (disprove) them. For the same reasons, I cannot deny God with certainty, which would make me an atheist. This is a conclusion reached by many scientists. I simply admit that there is so much that I don't know and that will always remain beyond my (and mankind's) comprehension. Fortunately, I have never had difficulty admitting my limitations (and there are many). Scientists, however, and particularly the more successful ones, are not always prepared to say that there is much they just don't know and that much will stay incomprehensible. In a way, they disregard Kurt Gödel's incompleteness theorem (according to which in mathematics, and thus probably in other sciences, there are insolvable problems) and believe science can eventually provide all the answers. They are consequently tempted to push for justification of their views, their theories, and their assumed proofs. The superstring theory is again an example. One day it indeed may succeed, combining all particles and forces into one complex mathematical equation of 11 dimensions. But what will be its real meaning? If there is a creator, was the creator really dealing with an 11-dimensional, highly complex mathematical system in designing the universe? If, on the other hand, there was no creator or higher intelligence and thus no predetermined design, was it, as Monod argued, only chance that eventually determined the emergence of our universe and our being? Many cosmologists believe in the anthropic principle according to which the universe is as it is because if it were not and only the slightest changes in physical laws had come about, there would be no intelligent life whereby the universe could be known. Knowledge, after all, is our perception of everything. Without the existence of intelligent life (as contrasted with lower, primitive forms of living organisms) there could be no quest for understanding of our physical and biological world. The same is true for the universe with its innumerable celestial bodies (or even a multiverse cosmos) and questions of how it all came about and what its purpose and destination are.

In different ways, Monod and Popper suggested that at the interface of reason (consciousness) and the brain a discontinuum must exist. Consciousness (reason) can direct physiological processes in the brain, which in a way denies the principle of conservation of energy because material effects will give further impulses, thus causing either the kinetic or potential energy to increase. The laws of thermodynamics in extreme cases also can no longer be valid. This may be the case for the "big bang" conditions of the initial state of the universe, in collapsed stars, or near black holes at enormous densities and pressures. At the border of mentality, a similar irregulatory (discontinuum) would exist. Thus the physical laws of our universe themselves cannot be considered truly universal. If there indeed are countless other universes (multiverses), their laws could be different, but they will remain inaccessible to mankind.

For me, it is not difficult to reconcile science (and by necessity our limited knowledge) and the possibility (although to me not probability) of a higher being or intelligence beyond our grasp and understanding. Some call it the reconcilability of science and religion. I would not, however, equate the consideration of a higher being or intelligence with religion. Religion is generally considered the practice of a belief in a divine power according to specific conduct or rules. Organized religions (Christianity, Islam, Judaism, etc.) have, for example, difficulties in accepting many scientific facts. For example, the Book of Genesis represents a timeline of about 6700 years since the "creation of Man." If we accept fossil evidence and other evidence of evolution (the Pope himself indicated recently that evolution is indeed probable), this timeline certainly cannot be taken verbatim. Of course the Bible's six days of creation may not represent the equivalent of today's earth days but, more probably, periods of possibly very long duration, even billions of years. In any case, evolution cannot, and never attempted to, answer the question of how life originally started. Darwin himself was not an atheist.

Monod's view that it was only chance that brought about life by forming essential building blocks from innumerable individual atoms is in contradiction with mathematical probability. Einstein said that "God is not playing with dice," and himself was a believer. However, it is not necessary to consider that random combination of atoms somehow, despite overwhelming mathematical improbability, resulted in life. We know now that certain essential building block organic molecules (including amino acids, nucleic acid bases, etc.) could be formed from basic inorganic molecules prevalent in the cosmos and containing only a few atoms. It is their combination (and not that of the random combination of all the atoms contained in them) that could have produced our complex biological systems. Spontaneous assembly of some fairly complex molecules using proper templates is now probed by chemists in their laboratories. This, however, would not represent creating life, certainly not intelligent life, in a test tube.

When considering how the evolution of life could have come about, the seeding of terrestrial life by extraterrestrial bacterial spores traveling through space (panspermia) deserves mention. Much is said about the possibility of some form of life on other planets, including Mars or more distant celestial bodies. Is it possible for some remnants of bacterial life, enclosed in a protective coat of rock dust, to have traveled enormous distances, staying dormant at the extremely low temperature of space and even surviving deadly radiation? The spore may be neither alive nor completely dead, and even after billions of years it could have an infinitesimal chance to reach a planet where liquid water could restart its life. Is this science fiction or a real possibility? We don't know. Around the turn of the twentieth century Svante Arrhenius (Nobel Prize in chemistry 1903) developed this theory in more detail. There was much recent excitement about claimed fossil bacterial remains in a Martian meteorite recovered from Antarctica (not since confirmed), but we have no definite proof for extraterrestrial life or its ability to travel through space. In the universe (or multiverse) there indeed may be many celestial bodies capable of maintaining some form of life (maybe entirely different from our own terrestrial forms), but will we ever be able to find out about them, not to mention communicate with any higher intelligent beings? This will probably remain for a long time only speculation for humankind.

Concerning intelligent life, *Homo sapiens* has been present on earth for only a short period of time of some tens or hundreds of thousands of years compared to the 4.6 billion-year history of our planet. Simple forms of living organisms were around for billions of years, but man's evolution was slow. At the same time, who can say how long we will be around? According to the theory of Darwinian evolution, species disappear and are replaced by more adept ones. For example, if the dinosaurs had not become extinct some 65 million years ago when an asteroid hit the earth, mankind probably could not have evolved. The natural selection process continues to go on, and human activities could even accelerate it. Environmentalists argue that all existing species must be preserved, even a tiny fish or a rare bird in remote areas. While we strive to maintain our environment as free from human influence as possible, natural processes inevitably will go on. A major catastrophic event, such as an asteroid colliding with earth, may one day make extinct many of the present life forms. Lower forms of life, such as bacteria, will in all probability survive, and the process of evolution could restart. The resulting higher species, however, may turn out to be different from those that we know today. This may be also the case if intelligent species exist in other parts of the immense cosmos, whose presence, however, we may never be able to ascertain or communicate with. Our limited biological life span, besides other factors, is an obvious limit for deeper space travel and adds to our limitation to "learn everything." We also must realize how fragile and short-lived mankind probably really is.

When considering the place of science in mankind's overall effort for knowledge and self-expression, it is striking to realize how much interrelationship exists between our different intellectual activities. Man's drive to express himself can take different forms. Some of these involve making or doing things such as painting, sculpture, architecture, music, writing, drama, dance, and other activities in what we call the arts and letters. The field of humanities in general is concerned with learning about and expressing human thought and relationships. The sciences, as discussed are concerned with knowledge and understanding of nature, the physical world, and the forces, laws, and rules governing them. For science to move forward to new levels of understanding, we need to advance creative new ideas, concepts, and theories and to explore their reality. In this sense, science in many ways is not much different from the concepts or thoughts of the humanities and letters or forms of self-expression in the arts, music, etc. Arts, humanities, and creative sciences are closely related, even if this is not always fully recognized. Of course, we must differentiate the artist from the artisan, the composer or creative musician from the mere practicing performer, as much as the "regular" scientist or technologist from the creative one ("revolutionary" in Thomas Kuhn's sense).

Recently, I have been teaching a freshman seminar on the relationship of the sciences with the humanities and economics. One semester of discussions, for example, led with an economist and a humanist colleague, was centered on Goethe's *Faust* as an example of the interrelationship between our seemingly unrelated fields. Faust is the epitome of Goethe's life experience. He was a great poet, but also a remarkable man with wide interest in different fields including the sciences and economics (he was for a while the finance minister of his German principality). The first part of the Faust story can be looked at as the story of an alchemist (i.e., early chemist) who strives through the philosopher's stone to make gold. Even as he fails, in the second part of the Faust story Goethe discusses paper money as a way achieving his goal. Paper money assumes the role of gold and even creates new capital and wealth. Goethe thus showed the interrelationship of economics with alchemy. The story of Faust also gives us a sense of the state of alchemy (i.e., early chemistry) in that period of history.

There are many other examples of interrelationship. "Symmetry," for example, is of fundamental importance in the sciences and arts alike. It plays a key role in our understanding of the atomic world as well as the cosmos. The handedness of molecules, with nature selecting one form over the other, contributes fundamentally to the evolution of living organisms. Symmetry also plays a significant role in typographical number theory in mathematics. It is also of great significance in the complex string theory through which mathematical physics is trying to develop a "theory for everything." At the same time there is a key role of symmetry in the arts expressed by varying examples from the fascinating paintings and graphics of Escher to Bach, who in some of his sonatas wrote two (or more) independent musical lines to be played simultaneously, in a way creating a musical symmetry effect.

As science in some way or other affects practically all aspects of life, without necessarily attempting to give a deeper understanding of its complexities, it is essential that all educated people in the modern world have at least a rudimentary education in science. Literacy should not only mean being able to read and write (or use a computer) but also having at least a minimal "science literacy." At all levels of science education, the clarity of presenting facts and concepts is of great importance but should not be at the expense of accuracy. This is not easy, because science should also be presented as a fascinating, dynamic, and challenging topic that should catch the attention of children and adults alike and inspire them to follow up with more detailed studies and reading.

The twentieth century was considered the century of science and technology. It produced many renowned scientists, some of whom, such as Albert Einstein, gained wide general recognition. Science education at the same time in the post-Sputnik second half of the century started to lose some of its shine and cultural significance. It is difficult, however, to imagine how tall edifices can be built without proper foundations. The increasing interest in interdisciplinary studies frequently also puts premature emphasis on crossing different disciplines without first establishing solid foundations in them. Many learn readily the vocabulary and superficial aspects of a field but lack solid grounding and knowledge. There is the danger that the underpinnings essential for science will be weakened. The trailblazers of the DNA revolution, Crick and Watson, are household names of twentieth-century science. At the same time, nature does not readily or frequently give such intuition and recognition. It is necessary to be well prepared and able to pursue the science of frequently broad and complex fields with all the skills and knowledge acquired through solid education. In my field of chemistry, "magic" certainly comes very rarely and even then generally only coupled with consistent, hard work and study. Ideas, of course, are the essence of new discoveries, but at the same time one must be well prepared to realize which has merit and significance, as well as be able to stay the course to follow them through. There is usually little glamor in science compared with the long and frequently disappointing efforts it demands. There is, however, the occasional epiphany of discovery and fundamental new understanding, the eureka or ecstasy that makes it all worthwhile, but this is something only those who have experienced it can really appreciate. I may be in some small way one of the lucky ones.

· 3 ·

Chemistry:
The Multifaceted Central Science

As a chemist, I should briefly discuss what my field of science is. Here I also reflect on its historical development and scope, which help to put in perspective the broad background on which our contemporary chemistry was built, and where my own work fits in.

Chemistry deals with substances, their formation, and subsequent transformations as well as their composition, structure, and properties. Chemistry does not deal with either the infinitely small world of subatomic particles or the cosmological mysteries of the infinitely large cosmos, although extraterrestrial chemistry is involved in the material universe. Chemistry does not directly deal with the living world, but it is essential for our continued understanding of the world at the molecular level. Chemistry is thus also essential to our understanding of other sciences and is recognized to be the central science bridging physics and biology, drawing on the basic principles of physics while enabling us to understand biological systems and processes at the molecular level.

The concepts of chemistry were formulated on the observation and study of various elements and their compounds. Matter was suggested to be composed of indivisible particles called atoms by the ancient Greek philosophers. In modern times more than 100 different kinds of atoms are recognized, composing the chemical elements. When atoms combine, they form molecules and compounds (an assembly of a large number of molecules). They are held together by forces generally referred to as chemical bonding. In the strict sense, no such thing as the "chemical bond" exists, only atoms held together by sharing electrons in some way (covalent bonding) or by electrostatic charge attractions (ionic bonding). The highest probability of the location of electrons between atoms is depicted by the chemist by two-, three-, or multicenter bonding (sharing electrons). The transformation of molecules and compounds by various changes leads to new and different molecules.

Because of chemistry's very wide scope, it is customary to divide it into branches. One of the main branches is *organic chemistry*, which originally dealt with compounds that were obtained from (or related to) living organisms but is now generally recognized to be the chemistry of the compounds of carbon or, more precisely, of hydrocarbons (compounds of carbon and hydrogen) and their derivatives. *Inorganic chemistry* deals with compounds of the elements other than organic compounds (i.e., hydrocarbons and their derivatives). In *biochemistry*, the compounds and chemical reactions involved in processes of living systems are studied. *Biological chemistry* (more recently also *chemical biology*) involves the chemistry of biological systems. *Physical chemistry* deals with the structure of compounds and materials as well as the energetics and dynamics of chemical changes and reactions. It also includes related theoretical studies (*theoretical chemistry*). *Analytical chemistry* encompasses the identification and characterization of chemical substances as well as their separation (isolation) from mixtures.

There is an ever-increasing number of further subdivisions, or what I would call "hyphenated" branches of chemistry or chemically related sciences. Whether "chemical-physics" or "chemical-biology" is more meaningful than "physical chemistry" or "biological chemistry" may depend on the point of view one wants to look from.

Chemistry is the science of molecules and materials, physics deals with forces, energy, and matter (also including fundamental questions of their origin), and biology deals with living systems. Chemistry deals with how atoms (formed from the original energy of the big bang) build up molecules and compounds, which eventually organized themselves into more complex systems of the physical and biological world. It also deals with man-made compounds and materials. Chemistry is not directly concerned with such fundamental questions as how the universe was formed, what (if any) the origin of the big bang was, what the nature of the infinite minuscule subatomic world is, or, on the other hand, the dimensionless cosmos, how intelligent life evolved, etc. It deals with molecules composed of atoms of the elements and their assembly into materials or biological systems with their eventual enormous complexity. It is frequently said that in the unofficial order of the sciences physics comes first (mathematics is not considered strictly a science but rather a way of expression of human knowledge, not unlike language), followed by chemistry and then biology. This is also, incidentally, the sequence in which the prizes are presented during the Nobel ceremony, although there is no prize for biology as such, only medicine or physiology (more about this in Chapter 11).

What we now call chemistry slowly emerged over the centuries as mankind's use of varied substances and compounds and the quest for understanding of the material world evolved. The practical beginning of chemistry goes back to ancient Egypt, based on experience gained in metals, glass, pottery, tanning and dying substances, etc. On the other hand, speculations by the Greeks and peoples in the East laid the foundation of this quest for a better understanding into the nature of the material world.

It was in the great school of Alexandria that these separate paths came together and eventually led to the alchemy and iatrochemistry of future generations and, eventually, the chemistry of modern science.

In all the early natural philosophies, there is the underlying idea that there was some primordial element or principle from which the universe was derived. It was perhaps Thales who in his doctrine first speculated that water was the prime element. Plato, in his Timotheus based on Aristotle, suggested that four elements made up all things in the universe: earth, water, air, and fire. These platonic elements were assigned characteristic geometric shapes. The elements were mutually transformable by breaking down their geometric shapes into those of the others. The doctrine of the four elements was taught by Aristotle, who emphasized the broad principle that one kind of matter can be changed into another kind; that is, transmutation is possible. Aristotle's concept differed fundamentally from that of the unchangeable elements (Empedocles) and the mechanical hypothesis of Democritus, according to which the world was built upon the meeting of rapidly moving atoms, which themselves, however, are of unalterable nature. In Egypt and the area of Mesopotamia where working of metals was advanced, these concepts gained roots. When the Arabs conquered Egypt in the seventh century and overran Syria and Persia, they brought a new spirit of inquiry onto the old civilization they subdued.

To the development of what eventually emerged as the science of chemistry, metallurgy and medicine made contributions as well, but these origins have not crystallized into a unified picture. The fourth major contributor was alchemy, which originated in Egypt and the Middle East and had a twofold aspect. One aspect was practical, aimed at making gold from common base metals or mercury and thus providing unlimited wealth for those who could achieve it. The other aspect was the search in the medieval world for a deeper meaning between man and the universe and of general knowledge based on the elusive "philosopher's stone."

These days we consider alchemy a strange and mystical mixture of magic and religion, at best an embryonic form of chemistry but more a pseudo-science. But as Jung pointed out, alchemy was not simply a futile quest to transform base metals into noble gold. It was an effort in a way to "purify the ignoble and imperfect human soul and raise it to its highest and noblest state." It was thus in a way a religious quest —not necessarily just a scientific one. Matter and spirit were inseparable to medieval alchemists, and they strove to transform them through these procedures, which sometimes amounted to sacramental rites and religious rituals as much as scientific research.

This is not the place to discuss the frequently reviewed historical and philosophical aspects of alchemy, but it is worthwhile to recall some rather late adherence to the precepts of alchemy by giants of the human intellectual endeavor. Johann Wolfgang Goethe is best known for his poetry and literature as the author of *Faust*. He himself, however, considered some of his major achievements to be in science. His interests were varied but also related to chemistry. He developed an early interest in alchemy, which, however, he overcame in later life. Goethe's classic character Faust reflects his fascination with the alchemist's effort to produce gold but eventually recognizes its futility and failure.

Newton, one of the greatest physicists of all time, is said to have spent more time on his alchemist efforts and experimentation than on his physical studies. Was this only scientific fascination, or had his position as Master of the Royal Mint added to his interest? Not only could he have wanted to safeguard the purity of the gold coins of the realm, but the thought may have crossed his mind that perhaps he could also produce the gold itself by transforming much less valuable mercury. It is, however, also possible that he was looking for an understanding of the elemental matter and possible transformation of elements. Newton also failed in his quest, and it was only in our nuclear age that transmutation of the elements was achieved. In fact, in 1980 a bismuth sample was transmuted into a tiny amount of gold in the Lawrence–Berkeley Laboratory, although at a very high cost.

In the narrow sense of the word, alchemy is the pretended art of transmuting base metals such as mercury into the noble ones (gold and silver). Its realization was the goal up to the time of Paracelsus and even later. Alchemy in its wider meaning, however, stands for the chemistry of the middle ages. Alchemy thus in a sense focused and unified varied and diverse chemical efforts, which until that time were disconnected, and focused them on producing varied practical materials for human needs. Alchemy indeed can be considered an early phase of the development of systematic chemistry. As Liebig said, alchemy was "never at any time anything different from chemistry."

It also must be noted that the processes described by the alchemists going back to the thirteenth century were generally not considered to be miraculous or supernatural. They believed that the transmutation of base metals into gold could be achieved by their "art" in the laboratory. But even among the late Arabian alchemists, it was doubted whether the resources of the art were adequate to the task. In the West, Vincent of Beauvais already remarked that success had not been achieved in making artificial metals identical with the natural ones. Roger Bacon, however, still claimed that with a certain amount of the "philosopher's stone" he could transmute a million times as much base metal into gold.

In the earlier part of the sixteenth century Paracelsus gave a new direction to alchemy by declaring that its true object was not the making of gold but the preparation of medicines. This union of chemistry with medicine was one characteristic goal of iatrochemists, of whom he was the predecessor. The search for the elixir of life had usually gone hand in hand with the quest for the philosopher's stone. Increasing attention was paid to the investigation of the properties of substances and of their effects on the human body. Evolving chemistry profited by the fact that it attracted men who possessed the highest scientific knowledge of the time. Still, their belief in the possibility of transmutation remained until the time of Robert Boyle.

It was indeed in the seventeenth century that chemistry slowly emerged in its own right as a science. Robert Boyle probably more than anybody else paved the way, helping to disperse its reputation as a tainted alchemical pseudo-science. In his book, *The Sceptical Chymist*, he emphasized the need to obtain a substantial body of experimental observation and stressed the importance for the quantitative study of chemical changes. Boyle is remembered for establishing that the volume of gas is inversely proportional to its pressure and for his pioneering experiments on combustion and calcination. He pointed out the importance of working with pure, homogeneous substances, and thus in a way he formulated the definition, but not necessarily the concept, of chemical elements, which he believed to be "primitive and simple, or perfectly unmingled bodies, not being made of any other bodies." However, he still believed that water, air, and fire were elementary substances. Nonetheless, Boyle believed in the atomic theory and that chemical combination occurs between the elementary particles. He also had good ideas about chemical affinity. His followers (Hooke, Mayow, et al.) extended his work.

Boyle and his followers represented the English school of chemistry, which, however, declined by the end of the seventeenth century, leading to the revival of the German iatrochemical school and introduction of the phlogiston theory. Iatrochemists believed that chemical substances contained three essential substances: sulfur (the principle of inflammability), mercury (the principle of fluidity and volatility), and salt (the principle of inertness and fixity). Becker, around the end of the seventeenth century, modified these three general constituents to represent terra lapida corresponding to fixed earth present in all solids (i.e., the salt constituent of the iatrochemists), terra pinguis, an oily earth present in all combustible materials (i.e., the sulfur constituent), and terra mercurialis, the fluid earth corresponding to the mercury constituent.

Stahl subsequently renamed the terra pinguis "phlogiston," the "motion of fire" (or heat), the essential element of all combustible materials. Thus the phlogiston theory was born to explain all combustion and was widely accepted for most of the eighteenth century by, among others, such luminaries of chemistry as Joseph Priestley.

Isolation of gases from calcination of certain minerals (fixed air) and from the air itself represented the next great advance in chemistry. Cavendish studied the preparation of hydrogen, "inflammable air," as he termed it. Priestley in the 1770s discovered and isolated several gases, namely, ammonia, hydrochloric acid gas, various nitrogen oxides, carbon monoxide, sulfur dioxide, and, most notably, oxygen, which he considered "dephlogisticated air." Independently, and even somewhat before Priestley, the Swedish apothecary Scheele discovered oxygen and pointed out that air could not be an elementary substance as it was composed of two gases, "fire air" or oxygen and "foul air" or nitrogen, according to a ratio of one to three parts by his estimate. However, Scheele still believed in the phlogiston theory, and he thought that oxygen's role was only to take up the phlogiston given out by burning substances. It was Antoine Lavoisier in France working in the latter part of the eighteenth century along rather different lines who systematically criticized the prevailing traditional chemical theories of his time. He realized that Priestley's "dephlogisticated air" was the active constituent of air in which candles burned and animals lived. Metals absorbed it on calcination. In 1775 Lavoisier thought that oxygen was the pure element of air itself, free from "impurities" that normally contaminated it. Scheele, however, showed in 1777 that air consisted of two gases, oxygen (which supported combustion) and nitrogen (which was inert). Lavoisier accepted Scheele's view and suggested in the next year that the atmosphere was composed of one-quarter oxygen and three-quarters nitrogen, a ratio that was subsequently corrected by Priestley to one-fifth oxygen and four-fifths nitrogen. Clearly, the discovery of oxygen and its role in combustion played a most significant role in the development of chemistry.

In 1783 Lavoisier announced a basic reevaluation of the "chemical theory," rejecting the phlogiston theory completely. At the same time, he elevated oxygen to a general explanatory principle (in a manner reminiscent of the iatrochemists), ascribing to it properties that were not experimentally warranted. For example, he suggested that oxygen was the basis of the acidifying principle, all acids being composed of oxygen united to a nonmetallic substance. This was subsequently disproved by Humphrey Davy, who in 1810 showed that hydrochloric acid did not contain oxygen. Lavoisier himself followed up his studies on combustion, discrediting the phlogiston theory with an effort to put chemistry on firm ground by suggesting an entirely new nomenclature to bring the definitions derived from increasing experimental facts into the general context of chemistry. The *Méthode de Nomenclature Chimique*, published in 1787, introduced names for 33 "simple substances" (i.e., elements) including oxygen, nitrogen (azote), hydrogen, etc., named acids, and their derived substances, i.e., salts. Through this new naming system (much of it still in use) Lavoisier also put the principles of chemistry on which it was based into common practice. He followed up with his famous *Traité Élementaire de Chimie*, a book that broke with traditional treatises. He discussed chemistry based on ideas backed by facts proceeding according to "natural logic" from the simple to the complex. He presented chemistry on the basis of analytical logic. He also broke with traditional historical pedagogy, according to which the three realms of nature were: mineral, vegetable, and animal. Lavoisier's treatise was the first modern work of chemistry and his major achievement in the "chemical revolution" he started. It is an irony of fate that this revolutionary chemist some years later lost his life to the guillotine of the French Revolution.

The dawn of the nineteenth century saw a drastic shift from the dominance of French chemistry to first English-, and, later, German-influenced chemistry. Lavoisier's dualistic views of chemical composition and his explanation of combustion and acidity were landmarks but hardly made chemistry an exact science. Chemistry remained in the nineteenth century basically qualitative in its nature. Despite the Newtonian dream of quantifying the forces of attraction between chemical substances and compiling a table of chemical affinity, no quantitative generalization emerged. It was Dalton's chemical atomic theory and the laws of chemical combination explained by it that made chemistry an exact science.

In 1804 Dalton formulated the concept that identified chemical elements with atoms. The notion of atoms, the smallest corpuscles of matter, was not new, of course, and had been around in some form or other since antiquity. Dalton, however, addressed the question to differentiate atoms not only by size (or shape) but also by their weight. To do this, Dalton turned to Proust's law, according to which the relationship of masses according to which two or more elements combine is fixed and not susceptible to continuous variation, and made it the center of his atomic hypothesis. He suggested that chemical combination takes place via discrete units, atom by atom, and that atoms of each element are identical. He also added the concept of multiple proportions; that is, when two elements form different compounds the weights in which one element will combine with another are in a simple numerical ratio. Dalton's atomic concept gave the whole body of available chemical information an immediate, easily recognizable meaning. What was also needed, however, was to relate all the atomic weights to a single unit. Dalton chose the atomic weight of hydrogen for this unit. Dalton's atoms also differed fundamentally from Newtonian corpuscles because they were not derived from an attempt to be based on the laws of motion and the attraction of single bodies whose ultimate constituents would be atoms.

Shortly after publication of Dalton's *New System of Chemical Philosophy* Gay-Lussac announced his observations that "volumes of gas which combine with each other and the volume of the combination thus formed are in direct proportion to the sum of the volumes of the constituent gases." The volumetric proportions of Gay-Lussac and Dalton's gravimetric ratios indeed supplement each other, although they themselves contested and rejected each other's concepts.

Whereas most chemists focused their attention on speculation about atoms and the question of atomic weights, the constant multiplicity in compounds occupied an increasingly central role. The new concept of substitution, i.e., the replacement of one element by another in a compound, started to make a major impact on chemistry in the 1840s. It was probably Dumas, who in the 1830s at the request of his father-in-law (who was the director of the famous Royal Sèvres porcelain factory) resolved an event that upset a royal dinner party at the Tuilleries in Paris. During the dinner, the candles in the hall started to emit irritating vapors, forcing the king and his guests into the gardens. Dumas found that the vapors were hydrochloric acid, which resulted from the replacement of hydrogen atoms by chlorine atoms during the bleaching of the wax of which the candles were made and subsequent decomposition. Dumas related this to other previous observations (by Faraday, Gay-Lussac, Liebig, and Wöhler) and his own study of the chlorination of acetic acid, establishing the principle of substitution.

The development of chemistry in the rest of the nineteenth century saw the emergence in Germany of such dominant persons as Wöhler and Liebig, who, helped by the work of the Swedish chemist Berzelius, spearheaded the emergence of organic chemistry. In 1828, Wöhler prepared urea by the rearrangement of ammonium cyanate (an inorganic salt) in aqueous solution upon heating. Up to this time, "organic" compounds had been obtained only from living organisms. Wöhler and Liebig's work preparing a whole series of these compounds by combining organic and inorganic components fundamentally changed this situation. Organic chemistry, i.e., the chemistry of carbon compounds, emerged and gained ever-increasing significance.

Liebig's students followed and greatly extended this new trend. My purpose here is not to trace the history of organic chemistry in any comprehensive way. Giants such as Bunsen, Kolbe, Baeyer, and Emil Fischer built a powerful tradition of synthetic and structural organic chemistry in the nineteenth century but at the same time tended to show significant antipathy toward any particular theory, despite the fact that physical chemistry created at the same time by Oswald and others made great progress.

Liebig's student Kekulé, however, was a most significant exception. His systematic classification of organic compounds led him to the realization of carbon's attribute of combining with four atoms or groups, just as oxygen binds to two atoms or groups. However, it probably was Frankland who first recognized the "saturation capacities of elementary atoms and the abilities of polyvalent atoms to couple in characteristic ways." Couper as well as Butlerov did much to point the way to express logical structural formulas. It was, however, Kekulé who in the 1860s developed more fully the concept of valence based substantially on his realization of the valence of four of carbon in its compounds. It was a century later that I was able to show that carbon in some systems can also bind simultaneously five, six, or even seven atoms or groups, introducing the concept of hypercarbon chemistry (see Chapter 10).

With the continued discovery of new elements in the nineteenth century, chemists started to group the elements together according to some empirical orders, which eventually developed into the Mendeleev periodic table. Mendeleev published his text, *Principles of Chemistry*, around 1870. Previously, Gerhardt had ordered most abundant chemistry around three typical molecules, H_2, H_2O, and NH_3, to which Kekulé added a fourth, CH_4. These molecules were seen to show a scale of equivalent valence ranging from one to four (i.e., H^IH, $O^{II}H_2$, $N^{III}H_3$, and $C^{IV}H_4$). The "typical" elements contained in this classification and their valences were well suited to a logical order of presentation. Because Gerhardt also included HCl as a typical molecule and this acid readily formed salts with alkali metals, Mendeleev in his first volume of *Principles* treated H, O, N, and C together with the four known halogens and alkali metals. Mendeleev's typical elements were all of low atomic weight and widely distributed in nature, thus being of organic type and not representing all elements. In his second volume, he added metals and began to order the elements according to their valence and atomic weight. The horizontal relationships related to valence, with a broad "transition" group following the alkali metals. Mendeleev was able to slowly fill gaps in his system, whereas other predicted "vacancies" were subsequently filled by discovery of new elements.

For nearly half a century, Mendeleev's periodic table remained an empirical compilation of the relationship of the elements. Only after the first atomic model was developed by the physicists of the early twentieth century, which took form in Bohr's model, was it possible to reconcile the involved general concepts with the specificity of the chemical elements. Bohr indeed expanded Rutherford's model of the atom, which tried to connect the chemical specificity of the elements grouped in Mendeleev's table with the behavior of electrons spinning around the nucleus. Bohr hit upon the idea that Mendeleev's periodicity could be explained by the limited number of electrons occupying the same orbital. When the orbital is filled, one moves down a line in the table.

Chemists were quick to appreciate Bohr's model because it provided an extremely clear and simple interpretation of chemistry. It explained the reason behind Mendeleev's table, that the position of each element in the table is nothing other than the number of electrons in the atom of the element, which, of course, represents an equal number of periodic changes in the nucleus. Each subsequent atom has one more electron, and the periodic valence changes reflect the successive filling of the orbital. Bohr's model also provided a simple basis for the electronic theory of valence.

Physicists traditionally pay little attention to the efforts of chemists. One can only wonder whether, if physicists had given more consideration to the chemists' periodic table, they could have advanced their concept of atomic theory by half a century! This is the irony of divided science, and it shows that chemistry is not a simple derivative of physics. It should be emphasized that all of science represents one entity. The understanding of the laws and principles of physics provides the foundation of our understanding not only of the limitless universe but also of the small elemental particles, the atoms that are the building blocks of the molecules and compounds of the chemist. These laws and principles also explain the forces that combine atoms (referred by chemists as chemical bonds) as well as the physical basis of valence. After Dirac derived his relativistic quantum mechanical equation for the electron as well as the physical basis of valence in 1928, he was said to have remarked that these, together with Schrödinger's equation describing H_2, explained all of chemistry, which can be simply derived from the first principles of physics. What he did not emphasize was that similar treatment from first principles of more complex, large molecules and those of nature alike would probably remain a challenge for a very long time.

For two thousand years atoms were considered the smallest and indivisible units of nature. At the beginning of the nineteenth century Dalton got chemistry on the path of atomic theory with his book, *A New System of Chemical Philosophy*, in which he argued that unbreakable atoms form compounds by linking with other atoms in simple whole-number proportions. It was at the end of the nineteenth century that Thompson discovered the existence of particles a thousand times smaller than the smallest atom. When it was found that these particles carry negative charge and are the fundamental unit of electricity, they were called electrons. Because, however, under normal circumstances atoms have no overall charge, there had to be a positive electrical component to neutralize the electrons. Rutherford subsequently showed that the center of the atom is a very dense nucleus that carries a positive charge. The negative electrons are orbiting, at substantial distance, the otherwise empty space of the atom. Rutherford's dynamic orbiting model, however, was unable to answer the question of why the moving charged particles (electrons), according to classical requirements of physics, do not lose energy and collapse into the nucleus. It was Bohr who subsequently linked Rutherford's atom with Planck's concept that energy, not unlike heat or light, is not continuous as Newton thought but exists in discrete quanta. Bohr developed his model of the atom in which electrons travel around the nucleus in circular orbits but only certain-sized orbits are possible, and these are determined by quantum rules.

The development of the structural theory of the atom was the result of advances made by physics. In the 1920s, the physical chemist Langmuir (Nobel Prize in chemistry 1932) wrote, "The problem of the structure of atoms has been attacked mainly by physicists who have given little consideration to the chemical properties which must be explained by a theory of atomic structure. The vast store of knowledge of chemical properties and relationship, such as summarized by the Periodic Table, should serve as a better foundation for a theory of atomic structure than the relativity meager experimental data along purely physical lines."

At the heart of chemistry, as Langmuir pointed out, was the periodic table. As mentioned since 1860, chemists had realized that although the elements could be arranged according to their increasing atomic weights, they contained properties that showed periodic similarities. The periods seemed to be eight elements long and then to repeat. The number eight seemed to be crucial, but no explanation was given until 1916, when G. N. Lewis explained this by his "octet rule," based on a cubelike model of the atoms, the electrons being equidistant from the nucleus. Moving up through the periodic system, new cubes were simply added.

Langmuir further developed the pioneering ideas of G. N. Lewis to bring chemistry in line with the new physics. Their combined concept also introduced the shared electron pair as the basis of chemical bonding. Sharing electrons was the glue, they said, that held molecules together. It is regrettable that Lewis, unlike Langmuir, did not receive the Nobel Prize in recognition of his fundamental contributions.

Lewis–Langmuir covalent electron pair bonding, combining two atoms by sharing two electrons (2e-2c bond) became a foundation of our understanding of chemical bonding. Two electrons, however, can be shared simultaneously not only by two atoms but, as shown in Lipscomb's extensive studies of boron compounds (Nobel Prize in chemistry, 1976), also by three atoms (2e-3c bond). It was my good fortune to realize that similar bonding is also possible in general with electron-deficient carbon (see Chapter 10). Kekulé's original concept that carbon can bind no more than four other atoms was thus extended into the new area of hypercarbon chemistry.

Chemistry as it was realized substantially derives from the interaction of electrons. The electronic theory of chemistry, particularly of organic chemistry, emerged, explaining the great richness of chemical observations and transformation, as expressed by Ingold, Robinson, Hammett, and many others following in their footsteps.

The advent of nuclear chemistry of the twentieth century not only allowed the creation of scores of new radioactive elements but also gave an explanation for how the formation of hydrogen and helium after the initial big bang was followed by the formation of the other elements. In accordance with Einstein's realization, energy equals matter (although terrestrial experimental verification was only recently obtained at the Stanford accelerator when it was shown that particles indeed are formed from high energy). According to the "big-bang" theory, all the matter in the universe initially was contained in a primeval nucleus of enormous density and temperature, which then exploded and distributed matter uniformly through space. As the temperature fell from an initial estimate of 10^{32} K after 1 sec to 10^{10} K, elemental particles (neutrons, protons, and electrons) formed. Upon further cooling, conditions became right to combine these particles into nuclei of hydrogen and helium: the process of element building began. From 10 to 500 sec after the big bang, the whole universe is thought to have behaved as an enormous nuclear fusion reactor. The composition of all atoms in the universe is suggested to be about 88.6% hydrogen, whereas helium makes up 11.3%. Together they account for 99.9% of the atoms.

When hydrogen is burned up in the nuclear furnace of a star, helium burning takes over, forming carbon, which in turn leads to oxygen, etc. Subsequent emission processes releasing α-particles, equilibrium processes, neutron absorption, proton capture, etc. lead to heavier elements.

Chemists are satisfied how atoms of the different elements could form from the initial enormous energy of the big bang explosion, without, however, the need to concern themselves with the reason for its origin. Atoms subsequently can combine into molecules, which in turn build increasingly complex systems and materials, including those of the living systems. This is the area of interest for chemists.

The second great advance of twentieth century physics came in the form of quantum theory. Quantum theory deals with probabilities. The great success of quantum mechanics and its wide applications cannot overcome this point. It was because of this that Einstein himself never fully accepted quantum mechanics (his frequently quoted saying was "God does not play dice"). He attempted for many years to develop a theory combining quantum mechanics and relativity but never succeeded. The effort still goes on for a unified theory, and it may one day succeed. As mentioned the string theory promises to develop a complex mathematical solution with some 11 dimensions, but it is unclear what new physical meaning will derive from it. For chemists the advent of quantum theory also brought new vistas. It was Pauling and subsequently others who introduced into chemistry the concepts of quantum mechanics. Pauling, for example, treated the structure of benzene in terms of what he called "resonance" (instead of "electron exchange" used by Heisenberg, Heither, London, and others). He assumed limiting forms to derive the more stable intermediate structure of higher probability, in a way mixing assumed single and double bonds in benzene, leading to stabilized partial double bonds with their peculiar properties. Pauling's resonance theory was much debated. In the 1950s it was strongly criticized in a pamphlet by Soviet scientists, with the approval of Stalin himself, which claimed that it was anti-materialist and anti-Marxist, because the limiting valance bond resonance "structures" had no physical reality. These critics, however, showed no understanding for the probability nature of quantum theory. (Some Russian scientists such as Syrkin and Dyatkina fully understood it, but were ostracized at the time for it.) When molecular orbital concepts were more widely introduced, the "controversy" faded away. Increasingly, the use of quantum mechanical calculational methods, first based only on approximations and later on more precise ab initio methods (Pople and Kohn, Nobel Prize 1998), allowed chemists (Schleyer and many others) to use calculational chemistry as a powerful tool to theoretically calculate not only energies but structures and expected reaction paths. Experimental chemistry, however, still remains the essence of chemistry, but computational methods greatly supplement it and in some cases point the way to new understanding and even unexpected chemistry. Organic chemistry in general made great strides. Synthetic methodology (Barton, Nobel Prize 1969), including the stepwise preparation of complex molecules (frequently those of natural products or biologically relevant systems), made and continues to make spectacular advances. Woodward's (Nobel Prize 1965) mastery of such syntheses is legendary, as is Corey's (Nobel Prize 1996) retrosynthetic approach, building molecules by reassembling them from their derived simpler building blocks. Asymmetric synthesis (i.e., that of chiral, optically active molecules) also advanced dramatically (Kagan, Nayori, Sharpless, and others). So did the synthesis and study of an ever-increasing array of inorganic compounds, including main group elements, organometallics (Fischer and Wilkinson, Nobel Prize 1973), metal-metal bound system (Cotton).

Chemistry also contributed in a major way to the development of modern biological sciences through an ever more sophisticated understanding at the molecular level. Long are gone the days when Emil Fischer, who can be credited as having established biochemistry through his pioneering chemical work on carbohydrates, amino acids, peptides, and his "lock and key" concept of enzyme action, practically gave it away, saying that proper chemists should not deal with compounds they cannot properly isolate, purify, crystallize, and analyze by the classical methods of his time. Chemistry has provided methods to isolate even the most complex bio-organic molecules, using a whole array of efficient separation and analytical (spectroscopic) methods, and in many cases even synthesize them, paving the way to molecular biology and biotechnology. X-ray crystallography of isolated and crystallized DNA (by Rosalind Franklin) made it possible for Crick and Watson (Nobel Prize in medicine 1972) to realize the double-helix structure of DNA (in contrast to Pauling's intuition of a triple helix, which in this case proved to be wrong).

I do not wish to go into further discussion of the only too well-known close interrelationship of chemistry and biology, which some these days like to call chemical biology instead of biological chemistry. The interface of chemistry and physics can be equally well called chemical physics or physical chemistry, depending on from which side one approaches the field. What is important to realize is that chemistry occupies a central role between physics and biology. Chemistry is a truly central, multifaceted science impacting in a fundamental way on other sciences, deriving as much as it contributes to them.

· 4 ·

Growing up in Hungary and Turning to Chemistry

I was born in Budapest, Hungary on May 22, 1927. My father, Gyula Olah, was a lawyer. My mother, Magda Krasznai, came from a family in the southern part of the country and fled to the capital, Budapest, at the end of World War I, when it became part of Yugoslavia.

Growing up in Budapest between the two World Wars in a middle-class family provided a rather pleasant childhood. Budapest was and still is a beautiful city. It is the only major city on the Danube through which the river flows through its middle in its true majesty. You can be in Vienna without necessarily realizing that it is also located on the Danube, as the river flows only through the outskirts. Not in Budapest! The half-mile-wide river divides Budapest into the hilly side (Buda) and the flat city (Pest). In fact, Pest and Buda were separate cities until 1873, when a permanent bridge was built to combine them. It is a beautiful chain bridge, and it still stands, after having been rebuilt following its destruction by the Nazis, together with all other bridges, at the end of World War II.

I was born in my parents' apartment in the Pest part of the city off the famous Andrassy Avenue and across from the Opera House. The apartment building at 15 Hajos Street was (and still is) an impressive building, designed by Miklos Ybl, a noted architect who designed the Opera House and other major public buildings. Many members of the Opera Company, from conductors to musicians to singers, were tenants in the same building. The composer Miklos Rozsa, who initially became famous for his musical scores for Alexander Korda's movies, grew up there, brought up by his uncle, who was an orchestra member.

I met him many years later in Los Angeles, and we reminisced about it.

Quite naturally, I attended many opera performances and rehearsals with neighbors and friends from early childhood on. I still remember a rehearsal conducted by a gentle, smallish man who became, however, a holy terror with the orchestra and the singers alike once he raised his baton. He was Arturo Toscanini.

I myself, however, had no musical inclination or talent. It was consequently somewhat amusing that years later, at the end of the war in the spring of 1945, friends made me a "Member of Budapest Opera House," allowing me to carry a rather official-looking identification card which was even respected by the occupying Soviet military and entitled me to some privileges. (Many civilians of all ages, including survivors of the Nazi terror, were taken indiscriminately into captivity as so-called "prisoners of war," frequently never to be heard from again.) My duties during my short-lived "operatic" career, however, included only such chores as clearing rubble or moving around pianos and other heavy musical instruments.

At the end of World War I, Hungary lost more than half of its former territory and population in the Versailles (Trianon) treaties. At the

With my parents and brother in 1928

My mother around 1943

same time, it became a much more homogenous small country of some 10 million people. The capital city Budapest retained its million inhabitants, acting not unlike Vienna as an oversized head for a much smaller body. Budapest also maintained an unusual concentration of industry and centralized bureaucracy and remained the cultural and educational center of the country.

Budapest between the two World Wars was a vibrant, cultured city with excellent theaters, concert halls, opera house, and museums. The city consisted of ten districts. The working-class industrial outskirts of Pest had their row-houses, whereas the middle-class inner city had quite imposing apartment buildings. The upper classes and aristocracy lived in their villas in the hills of Buda.

Some of the social separation, however, was mitigated in the schools. There were remarkably good schools in Budapest, although as with much else it cannot be said that there was a uniformly high level of education across the country. Much was written about the excellence of the gymnasia of Budapest (German-type composites of middle and high schools), particularly those that were frequented by some later successful scientists, artists, and musicians. For example, many Hungarian-born mathematicians, physicists, and engineers who later attained recognition (Neuman, Karman, Wigner, Szilard, Teller, among others) attended the same schools, where much emphasis was put on mathematics and physics. Winning the national competition in these topics for high school boys (the schools were still gender segregated and not much was ever said about the intellectual achievement of Hungarian girls, who clearly were at least as talented) was considered a predictor of outstanding future careers (Neuman, Teller).

I did not attend any of these schools, and there was, furthermore, no national competition in chemistry. After going to a public elementary school I attended the Gymnasium of the Piarist Fathers, a Roman Catholic religious teaching order, for eight years. Emphasis was on broad-based education, heavily emphasizing classics, history, languages, liberal arts, and even philosophy. The standards were generally high. Latin and German was compulsory for eight years and French (as a selective for Greek) for four years. During my high school years, at the recommendation of one of my teachers, I got my first teaching experience tutoring a middle school boy who had some difficulties with his grades. I enjoyed it and bought with my first earnings an Omega wristwatch, which I had for a quarter of century (I guess it was a good luck charm through some difficult times). I also took private lessons in French and English. In a small country such as Hungary, much emphasis was put on foreign languages. German was still very much a second language as a remnant of the Austro-Hungarian monarchy, and I was fluent in it from early childhood.

We also received a solid education in mathematics, as well as some physics and chemistry. Although, among others, the physicist Lorand Eotvos and the chemist George Hevesy (Nobel Prize in 1943) attended the same Piarist school in earlier years, I learned about this only years later and cannot remember that anybody mentioned them as role-models during my school years. Classes were from 8 AM to 1 PM six days a week, with extensive homework for the rest of the day. I did well throughout my school years in all subjects, except in physical education, where I remember I could never properly manage rope climbing and some exercises on gymnastic instruments. At the same time I was active and became reasonably good in different sports, including tennis, swimming, rowing, and soccer. In my time as a young faculty member at the Technical University I was the goalkeeper of the chemistry faculty team (a picture of which found its way onto the Swedish Academy Nobel poster published after my Prize).

My wife still teases me on occasion that I was always a stellar student (she uses the more contemporary expression "nerd") and school valedictorian. Studying, however, always came easy for me, and I enjoyed it. I was (and still am) an avid reader. In my formative school years I particularly enjoyed the classics, literature, and history, as well as, later on, philosophy. I believe obtaining a good general liberal

Goalie of the soccer team of the Chemistry faculty at the Technical University of Budapest (1952)

Not yet thinking about chemistry

education was a great advantage, because getting attached too early to a specific field or science frequently short-changes a balanced broad education. Although reading the classics in Latin at school age may not have been as fulfilling as it would have been at a more mature age, once I got interested in science I could hardly afford the time for such pleasurable diversions.

Besides the classics, Hungarian literature itself offered a wonderful treasure trove. It is regrettable that the works of many highly talented Hungarian poets and writers remained mostly inaccessible to the rest of the world because of the rather strange language in which they were written (Hungarian belongs to the so-called Finn-Ugoric family of languages, not related to any of the major languages). On the other hand, Western authors and poets were regularly translated into Hungarian, some of these translations being of extremely high quality. The Hungarian poet Janos Arany, for example, translated Shakespeare in a masterly way, although he taught himself English phonetically and could never speak it. I believe that many translations, such as ballads of Villon (one of my favorites), translated by the highly talented poet George Faludy, made them at least as enjoyable as the originals. Operas were sung only in Hungarian, which limited the chances for some talented Hungarian singers to perform in the West (which has changed since).

Western movies were subtitled, but an active domestic movie industry also existed. An indication of the talented people in this industry is reflected in the subsequent role Hungarian émigrés played in the movie industries of the United States and England. These were not limited to movie producers and executives (Adolf Zukor, Paramount Pictures; William Fox, 20th Century Fox; Sir Alexander Korda). Who would guess, considering the thickly accented English most Hungarians used to speak, that Leslie Howard's (Professor Higgins in *Pygmalion*) mother tongue was this rather obscure and difficult language? What a difference indeed from the charming accent of Zsa Zsa Gabor. On the wall of Zukor's office there was an inscription: "It is not enough to be Hungarian, you must be talented, too." Zukor was quoted to add in a low voice, "But it may help."

The standards of theatrical and musical life were also extremely high. Franz Liszt (a native Hungarian who achieved his fame as a virtuoso pianist and composer) established a Music Conservatory in Budapest around 1870, which turned out scores of highly talented graduates, including such composers as Kodaly and Bartok and conductors such as Doraty, Ormandy, Solti, and Szell.

Good teachers can have a great influence on their students. As I mentioned, besides the classics, humanities, history, and languages I also received a good education in mathematics and had some inspiring science teachers. I particularly remember my physics teacher, Jozsef Öveges, who I understand later introduced the first television science programs in Hungary, which made him well known and extremely popular. He was a very inspiring teacher who used simple but ingenious experiments to liven up his lectures. I must confess that I do not remember my chemistry teacher, who must have made less of an impression on me. When a friend received a chemistry set for Christmas one year, we started some experiments in the basement of his home. This was probably when I first experienced the excitement of some of the magic of chemistry. For example, we observed with fascination bubbles of carbon dioxide evolving when we dropped some sodium bicarbonate into vinegar or muriatic acid. Little did I expect to repeat this experiment with superacidic "magic acid" years later, when, unexpectedly, no bubbles were formed (see Chapter 8). After going through some of the routine experiments described in the manual that came with the chemistry set, we boldly advanced to more interesting "individual" experiments. The result was an explosion followed by a minor fire and much smoke, which destroyed the chemistry set and ended the welcome for our experimentation. I cannot remember thinking much about chemistry thereafter, until I entered university.

The outbreak of World War II in September 1939 initially affected our lives little. Hungary stayed out of the war until Hitler's invasion of the Soviet Union in June 1941, when it was pressured to join Germany, with disastrous consequences. In 1943, a whole Hungarian army corps was destroyed in Russia, and Hungary, not unlike Romania, subsequently tried to quit the war. In the spring of 1944, Hungary attempted the break with Germany but was unsuccessful, and the Gestapo with their Hungarian allies, the Arrow Crosses, took over and instituted a regime of terror. Much has been written about this horrible period, when hundreds of thousands perished, until World War II finally ended. The human mind mercifully is selective, and eventually we tend to remember only the more pleasant memories. I do not want to relive here in any detail some of my very difficult, even horrifying, experiences of this period, hiding out the last months of the war in Budapest. Suffice it to say that my parents and I survived. My only brother Peter (three years my senior), however, perished at the end of World War II in a Russian prisoner of war camp, where he was taken together with tens of thousands of civilians only to satisfy claimed numbers of prisoners of war. My home town, beautiful but badly ravaged Budapest, also survived and started recovering. However, soon again difficult times followed when a brief period of emerging democracy ended with the Communist takeover.

I graduated high school (gymnasium) right at the end of the war in the spring of 1945. We needed to pass a very tough two-day examination (matriculation or baccalaureate). This included, for example, translation of selected literature passages—without the help of a dictionary—not only from foreign languages (including Latin) into Hungarian, but also vice versa, not a minor effort. Having passed this

My brother Peter in 1943

barrier, it was time to think seriously about my future. It was clear that my interests in the humanities, particularly literature and history, did not offer much future in postwar Hungary. One inevitably was forced to think about more "practical" occupations. I had no interest in business, law, or medicine but was increasingly attracted to the sciences, which I always felt in a way to be closely related to the arts and humanities. My eventual choice of chemistry was thus not really out of character. I was attracted to chemistry probably more than anything else by its broad scope and the opportunities it seemed to offer. On one hand, chemistry is the key to understanding the fundamentals of biological processes and maybe life itself. It is also an important basis of the life and health sciences. On the other hand, chemists make compounds, including the man-made materials, pharmaceuticals, dyes, and fuels essential to everyday life. It also seemed to be a field in which, even in a poor, small country, one could find opportunities for the future.

My choice of chemistry as a career was from a practical point of view also not unusual. I read years later that Eugene Wigner, the

Hungarian-born physicist, when he was seventeen, was asked by his father (a businessman) what he wanted to do. The young Wigner replied that he had in mind to become a theoretical physicist. "And exactly how many jobs are there in Hungary for theoretical physicists?" his father inquired. "Perhaps three," replied young Wigner. The discussion ended there and Wigner was persuaded to study chemical engineering. Similarly, the mathematician John von Neuman initially also studied chemistry, although they both soon changed to their field of real interest.

I entered the Technical University of Budapest immediately after the end of World War II. It had been established some 150 years earlier as the Budapest Institute of Technology. Because the Austro-Hungarian education system closely followed the German example, the Technical University of Budapest developed along the lines of German technical universities. Many noted scientists of Hungarian origin such as Michael Polanyi, Denis Gabor (Nobel Prize in Physics 1971), George Hevesy (Nobel Prize in Chemistry 1943), Leo Szilard, Theodor von Karman, Eugene Wigner (Nobel Prize in Physics 1963), John von Neuman, Edward Teller, and others have studied there at one time or other, generally, however, only for short periods of time. Most completed their studies in Germany or Switzerland usually in physics or mathematics, and their career carried them subsequently to England or the United States. To my knowledge, I am the only one who got my entire university education at that institution and who subsequently was also on its faculty. I was proud to be recognized as such when my alma mater welcomed me back after many years in 1988 and gave me an honorary Doctor of Science degree.

Formally, chemistry was taught at the Technical University under the Faculty of Chemical Engineering, but it was really a solid chemical education (along the lines of related German technical universities), with a minimum of engineering courses, in sharp contrast to what American universities teach as chemical engineering. Classes were relatively small. We started with a class of about 80, but the number was rapidly pared down during the first year to half by rather demanding "do or die" examinations (including comprehensive oral examinations). Those who failed could not continue. This was a rather harsh process, but laboratory facilities were so limited that only a few could be accommodated. The laboratory training was thorough. For example, in the organic laboratory we did some 40 synthetic preparations (based on Gatterman's book). It certainly gave a solid foundation.

In contrast to the excellent general education some of the gymnasia of Budapest provided their students, the chemical education at the Technical University in retrospect was very one-sided. The emphasis was on memorizing a large amount of data and not so much on fostering understanding and critical probing. Empirical chemistry, including such fields as gravimetric analysis, the recounting of endless compounds and their reactions, was emphasized, with little attention given to the newer trends of chemistry developing in the Western countries. However, I was well prepared for self-study and gained much through long hours in the library, which opened up many new vistas. There were also other very positive aspects of my university education. Many of our professors through their lectures were able to induce in us a real fascination and love for chemistry, which I consider the most important heritage of my university days.

After completing my studies and thesis, in June 1949 I was named to a faculty position as an assistant professor in Zemplen's organic chemical institute. The following month I married my boyhood love.

I met Judy Lengyel, a shy girl not much interested in boys, in 1943 during a vacation at a summer resort not far from Budapest. I was 16, and she was 14. Little did I realize at the time that this would become the most significant and happiest event of my life. Although fairytale things are not supposed to happen in real life, we were married six years later and celebrated our golden wedding anniversary in 1999. If anybody was blessed with a happy marriage and life partnership, I am, there are no words that can express it adequately.

At the time of our marriage, Judy worked as a secretary at the Technical University. She subsequently studied chemistry and, after completing her studies, joined our research. To this day, however, I am not sure that she has completely forgiven me for what she probably rightly recollects was my enrolling her to study chemistry (taking advantage of my faculty position) without really getting her prior full approval. Even if I were guilty in this regard, my intentions were good. The life

The shy girl Judy Lengyel

As newlyweds in 1950

Under the oars held up by members of my rowing club after our wedding July 9, 1949

of any scientist is very demanding. If marriage is to be a real partnership, having a common understanding and interests in our professional life could help to make, or so I hoped, a marriage even more successful. I believe it worked out this way, but only through Judy's efforts and understanding. I can only hope that she indeed has forgiven me after all these years.

· 5 ·

Early Research and Teaching:
Departing the Shadow of Emil Fischer

I started my research in Budapest in the Organic Chemistry Institute of Professor Geza Zemplen, a noted carbohydrate chemist of his time and a student of Emil Fischer. Fischer was the towering organic chemist of the early twentieth century. His work, including that on amino acids and proteins, carbohydrates, the start of enzyme chemistry (his famous "lock and key" concept) and many other significant contributions, also laid the foundation for the explosive development of what became biochemistry and, in more recent times, molecular biology. He received the Nobel Prize in chemistry in 1902, the second year the prize was awarded. His students spread across not only Germany but the whole of Europe and established in their home lands the traditions of the Fischer school. They also contributed much to the study of nature's fundamental systems and their understanding. This was also true of Zemplen, who established in Hungary the first organic chemistry institute when he was named to a newly established chair in organic chemistry at the Technical University of Budapest in 1913. His work not only resulted in elevating the educational standards of his university together with the development of successful and productive research but also helped to build a successful and respected pharmaceutical chemical industry in Hungary, which continues to the present day.

Having brought home from Berlin Fischer's traditions and an overriding interest in natural products, particularly carbohydrate and glycoside chemistry, Zemplen established and ran his laboratory very much in the Fischer style, adding, however, his personal touches. The laboratories had tall ceilings and large windows (not unlike the Fischer laboratories which I have seen in the early 1950s at the Humboldt University in Berlin—to provide better air circulation because Fischer suffered from chronic hydrazine poisoning from his research). The fume hoods were vented by a gas burner providing draft through the heated air, certainly a not very efficient system by today's standards.

Zemplen, like Fischer, also ran his Institute in an autocratic way. Because the University was able to provide only an extremely meager budget, he purchased laboratory equipment, chemicals, books, etc. with his own earnings and consulting income. Like Fischer, Zemplen too expected his doctoral students to pay their own way. Fischer even charged them a substantial fee for the privilege of working in his laboratory. Becoming a research assistant to the professor was a great privilege, and, although it meant no remuneration, it also exempted one from any fees.

Zemplen was also a formidable character, and working for him was quite an experience, not only in chemistry. He liked, for example, "pubbing," and these events in neighborhood establishments could last for days. Certainly one's stamina and alcohol tolerance developed through these experiences. Recalling on occasion his Berlin days, Zemplen talked with great fondness of his lab mate, the Finnish chemist Komppa (of later camphene synthesis fame), who he credited not only with being a fine chemist but also with being the only one able to outlast him during such parties. In any case, none of this ever affected his university duties or his research.

Zemplen helped his students in many ways. I remember an occasion in the difficult postwar period. The production of the famous Hungarian salami, interrupted by the war, was just in the process of being restarted for export. The manufacturer wanted a supportive "analysis" from the well-known professor. Zemplen asked for a "suitable" sample of some hundreds of kilograms, on which the whole institute lived for weeks. When it was gone he rightly could offer an opinion that the product was "quite satisfactory." After the war, grain alcohol was for a long time the only available and widely used laboratory solvent, and, not unexpectedly, it also found other uses. Later, when it was "denatured" to prevent human consumption, we devised clever ways for its "purification." The lab also "manufactured" saccharine, which was then bartered for essential food staples in the countryside. One of my colleagues suffered serious burns one day when a flask containing chlorosulfuric acid broke in a water-ice bath, spraying acid into his face. Such was the price of moonlighting for survival.

Zemplen was a strong-minded individualist who opposed any totalitarian system, from Nazism to Communism. He was briefly jailed toward the end of World War II by the Hungarian Fascists for refusing to join in the evacuation of the Technical University to Germany when the Russian armies advanced on Budapest. He was also strongly opposed to the Communists.

As Zemplen was a student of Emil Fischer, therefore, I can consider myself Fischer's "scientific grandson." My initial research with Zemplen was along the lines of his interest in natural products and specifically centered on glycosides. Besides synthetic and some structural work, there also emerged practical aspects. Many glycosides present in nature have substantial pharmaceutical use. For example, some of the most effective heart medications were isolated from the pretty foxglove plant (*Digitalis purpurea*). Three cardiac agents had been prepared by the Swiss chemist Stoll, working for the Sandoz Company (digitoxin, gitoxin, and gitalin). Another member of the *Digitalis* family, *Digitalis lanata*, contains lanataglucoside C, whose aglucone is also digitoxin (still one of the most effective heart medications). As it turned out, the mild climate of the Tihany peninsula, jutting into famed Lake Balaton, favored the cultivation of this plant, and each year boxcar loads of its leaves were shipped to Sandoz in Switzerland. While working with Zemplen, I developed an improved process for the efficient isolation of lanataglycoside C, using treatment and solvent extraction of fresh leaves suppressing enzymatic degradation. This greatly increased yields, and a gram of pure lanataglycoside C could be obtained from a kilogram of leaves. Our process with Zemplen was patented in 1952 and, together with a process for partial hydrolysis to digitoxin, was successfully commercialized and used for years by the Richter Pharmaceutical Company. A booklet commemorating the 100th anniversary of the Hungarian Patent Office mentioned our patent as well as other Hungarian patents, such as those issued in the 1920s to Albert Einstein and Leo Szilard for an thermoelectric refrigerator (which, I understand, regrettably never brought any income to the inventors). In the 1930s in England, Szilard, the remarkable Hungarian-born physicist, applied for a patent on another invention, the principle of a thermonuclear device, the basis of the atomic bomb! In 1952, he and Enrico Fermi were issued an American patent for it, which by law the US Government acquired. What wide scope some inventors' creative minds cover!

I have fond memories of my brief period as a natural product chemist, particularly because it de facto still involved collecting and isolating products from nature's diverse plant life and doing chemistry directly on them. Most natural product chemists these days do not have this experience.

Around 1950, after having read about organic fluorine compounds, I became interested in them. I suggested to Zemplen that fluorine-containing carbohydrates might be of interest in glycoside-forming coupling reactions. I believed that selective synthesis of α- or β-glycosides could be achieved by treating either acetofluoroglucose (or other fluorinated carbohydrates) or their relatively stable, deacetylated free fluorohydrins with the appropriate aglucons. His reaction to my suggestion was, not unexpectedly, negative. To try to pursue organofluorine chemistry in postwar Hungary was indeed far fetched. Zemplen thought that the study of fluorine compounds, which necessitated "outrageous" reagents such as hydrogen fluoride, was foolish. It became increasingly clear that my ideas and interests did not match his. Eventually, however, I prevailed and, with some of my early dedicated associates, Attila Pavlath (who, after a career in industry and governmental laboratory research, at the time of this writing, had been elected the President of the American Chemical Society) and Steve Kuhn, started to study organic fluorides. Laboratory space and particularly fume hoods were scarce in the Zemplen laboratories, and even when I became an assistant professor I was not welcome to "pollute" space intended for "real chemistry." However, at the top floor of the chemistry building, overlooking the Danube and badly damaged during the war, was an open balcony used to store chemicals. In an unexpected gesture, Zemplen allowed me the use of this balcony. With some effort we enclosed it, installed two old fume hoods, and were soon in business in what was referred to as the "balcon laboratory." I am not sure that Zemplen ever set foot in it. We, however, enjoyed our modest new quarters and the implicit understanding that our fluorine chemistry and subsequent study of Friedel-Crafts reactions and their intermediates was now approved.

Carrying out fluorine research in postwar Hungary was clearly not easy. There was no access even to such basic chemicals as anhydrous HF, FSO_3H, or BF_3, and we had to prepare them ourselves. HF was prepared from fluorspar (CaF_2) and sulfuric acid, and its reaction with SO_3 (generated from oleum) gave FSO_3H. By treating boric acid with fluorosulfuric acid we made BF_3. The handling of these chemicals and their use in a laboratory equipped with the barest of necessities was indeed a challenge. Zemplen's consent to our work was even more remarkable, because he truly belonged to the "old school" of professors who did not easily change their mind. For example, he never believed in the electronic theory of organic chemistry. In 1952, he was persuaded to publish a Hungarian textbook of "Organic Chemistry," which contained a wealth of information and discussion of natural products but otherwise still treated organic chemistry only in the descriptive ways of the nineteenth and early twentieth centuries. An interesting side aspect of this book was that it contained a substantial index, for which, with a faculty colleague (Lajos Kisfaludy), we were commissioned by the state publishing house on a per page basis. Our rather precarious financial existence (our meager university salary amounted to the equivalent of less than \$50/month, which was gone by the middle of the month) inspired us to a remarkable effort. Our index eventually contained 250 pages (a fifth of the book) and even displayed the preparers' names. I believe it is the longest index of any textbook of chemistry. It had, however, seen us through some difficult times.

No mechanistic aspects of organic chemistry (or, for this reason, any reaction intermediates) were ever mentioned by Zemplen in his lectures or writings, nor did he consider or accept their existence. I never heard him mention the names of Meerwein, Ingold, Robinson, or any other pioneers of the mechanistic electronic theory of organic chemistry. The possible role of organic ions was similarly never mentioned. He was, however, not alone at that time in this attitude. Continental, particularly German chemistry generally was slow to accept Robinson and Ingold's electronic-mechanistic trends. Roger Adams, one of the leading American organic chemists of the time (and a descendent of the family that gave two presidents to America), for example, never believed in organic ions, and for a long time his influence prevented anybody (including Whitmore, who pioneered the concept of the role of cationic intermediates in organic chemistry in America) from publishing any article in the *Journal of the American Chemical Society* mentioning such ions. Clearly, working on such problems in the Zemplen Institute came close to sacrilege, but I somehow got away with it. Perhaps this was in part due to the fact that Zemplen's health was declining at the time in his long battle with cancer; he came less frequently to the Institute, working mostly at his nearby apartment.

Did the Technical University in Budapest at this time provide a reasonable education and research atmosphere in chemistry? In retrospect, it is difficult to answer this question because of the prevailing isolation and primitive working conditions. The scope and topics of major courses taught, with some exceptions, mainly demanded memorization of facts and not understanding and discussion of concepts. Laboratory experimental work and research were much emphasized, despite the very limited facilities and opportunities. The overall system, however, prepared students well for self-study. I myself discovered at an early age the joy of broadening my scope and knowledge during long hours spent in the library in extensive reading and self-study (a habit I have kept all my life). Professors always lectured themselves, which was considered a major prerequisite of the professorship. Attendance by all students was strictly obligatory (the system was tailored on the German example). Lectures by popular professors were considered worthy and were well-attended events. I remember also attending many lectures outside my field in the Budapest universities such as by philosophers historians, which were always packed to the rafters with interested students.

In chemistry, there was clearly a great need to move ahead and bridge the gap between the earlier, entirely empirical approach of teaching and research and that incorporating the new trends of chemistry. The emerging understanding of mechanistic-structural aspects, application of physical methods and considerations, and the basic principles of bonding, reactivity, etc. were rarely mentioned at the time. When as a young assistant professor I had a chance starting in 1953 to teach my own course, it was entitled "Theoretical Organic Chemistry" (although really it was mechanistic and structural, i.e., physical organic chemistry). It filled a gap for which I tried to introduce the evolving new frontiers of organic chemistry. I published a two-volume text based on my notes from the two-semester course given in 1953–1954. An extended and reworked first volume in German was submitted to the publisher in 1956 but was published only in 1960 (G. Olah, *Theoretische Organische Chemie*, Akademie Verlag, Berlin, 1960) because of events that led to my move to America. The second volume of the German edition was never published. At that time in Germany, physical organic chemistry was also slow to develop. There was an early prewar text by Eugen Müller, but it was Heinz Staab's book in the 1950s that opened up the field. Looking up my own book occasionally, I realize that despite my isolation beyond the Iron Curtain, I still somehow managed to get myself oriented in the right direction. My book at the time was reasonably up to date and in some aspects maybe even novel and original.

During my time on the faculty of the Technical University I was also involved in helping to start a technical high school associated with the university. This school also filled a gap, because it was set up to allow evening study for those who did not have the opportunity for a high school education and were working at the university and in the surrounding community. It provided solid secondary education with an emphasis on the sciences and technology, qualifying its students for meaningful employment as technicians and in other technically oriented jobs. I remember that I even taught physics for a year, when we had difficulty finding a teacher for physics. To my surprise, I managed it quite well.

My teaching experience was, however, only secondary to my research interest. Through my initial research work involving reactions of fluorinated carbohydrates I became interested in Friedel-Crafts acylation and subsequently alkylation reactions with acyl or alkyl fluorides, with boron trifluoride used to catalyze the reactions. This also was the beginning of my long-standing interest in electrophilic aromatic and later aliphatic substitution reactions.

$$ArH + RCOF \xrightarrow{BF_3} ArCOR + HF$$

$$ArH + RF \xrightarrow{BF_3} ArR + HF$$

These studies at the same time aroused my interest in the mechanistic aspects of the reactions, including the complexes of RCOF and RF with BF_3 (and eventually with other Lewis acid fluorides) as well as the complexes they formed with aromatics. I isolated for the first time at low temperatures arenium tetrafluoroborates (the elusive σ-complexes of aromatic substitutions), although I had no means to pursue their structural study. Thus my long fascination with the chemistry of carbocationic complexes began.

$$ArH + HF + BF_3 \rightleftharpoons ArH_2^+ BF_4^-$$

$$ArH + RF + BF_3 \rightleftharpoons ArHR^+ BF_4^-$$

My early work with acyl fluorides also involved formyl fluoride, HCOF, the only stable acyl halide of formic acid, which was first made in 1933 by Nyesmeyanov, who did not, however, pursue its chemistry. I developed its use as a formylating agent and also explored formylation reactions with CO and HF, catalyzed by BF_3.

Another aspect of my early research in Budapest was in nitration chemistry, specifically the preparation of nitronium tetrafluoroborate, a stable nitronium salt. I was able to prepare the salt in a simple and efficient way from nitric acid, hydrogen fluoride, and boron trifluoride.

$$HNO_3 + HF + 2\,BF_3 \longrightarrow NO_2^+BF_4^- + BF_3.H_2O$$

This salt turned out to be remarkably stable and a powerful, convenient nitrating agent for a wide variety of aromatics (and later also aliphatics). Over the years, this chemistry was further developed, and nitronium tetrafluoroborate is still a widely used commercially available nitrating agent.

$$ArH + NO_2^+BF_4^- \longrightarrow ArNO_2$$

$$RNH_2 + NO_2^+BF_4^- \longrightarrow RNHNO_2$$

$$ROH + NO_2^+BF_4^- \longrightarrow RONO_2$$

In the course of my studies I also introduced silver tetrafluoroborate, $AgBF_4$, as a metathetic cation forming agent suitable for forming varied ionic (electrophilic) reagents.

$$RCl + AgBF_4 \longrightarrow R^+BF_4^- + AgCl$$

My publications from Hungary in the early 1950s somehow caught the eye of Hans Meerwein, a towering and pioneering German chemist. It is still a mystery to me how he came to read them. In any case I received an encouraging letter from him, and we started corresponding (even this was not easy at the time in a completely isolated Hungary). He must have sympathized with my difficulties, because one day through his efforts I received as a gift a cylinder of boron trifluoride. What a precious gift it was indeed, because it freed us from laboriously making BF_3 ourselves.

My early research on organofluorine compounds starting in 1950 centered on new methods of their preparation and use in varied reactions. At the time I also started a collaboration with Camillo Sellei, a wonderful physician associated with the Medical University of Budapest, on the study of the pharmacological effect of organofluorine compounds and particularly on the use of some organic fluorine compounds in cancer research. Sellei and his researcher wife Gabi became close friends and the godparents of our older son, George. Our joint publications from 1952 to 1953 must be among the earliest ones in this intriguing field. In subsequent years, the work of Charles Heidelberger (who became a colleague of mine when I moved to the University of Southern California in Los Angeles) and other researchers introduced such widely used fluorinated anticancer drugs as 5-fluorouracil. The pursuit of the biological activity of fluorinated organic compounds clearly was a worthwhile early effort, as progress in this still rapidly expanding field was and still is remarkable.

At the time, I was so fascinated by the potential of our organofluorine compound-based pharmaceutical research that, to learn more about biological aspects and medicine while maintaining my university appointment and duties, I enrolled at the Medical University. I passed all the preclinical courses, including the rather challenging anatomy course, but never really intended to obtain an MD. My increasingly active research interest and teaching duties led me after two years to abandon further medical education. What I learned, however, served me well, because it is difficult to appreciate the biomedical field and its revolutionary expansion without some proper foundations.

The Hungarian educational system after the Communist takeover was realigned in the early 1950s according to the Soviet example. University research was de-emphasized, and research institutes were established under the auspices of the Academy of Sciences. I was invited in 1954 to join the newly established Central Chemical Research Institute of the Hungarian Academy of Sciences, which was headed by Professor Geza Schay, a physicochemist with interest in surface chemistry and catalysis. He was a warm, truly fine man, who considered it his main goal to help the development of his younger colleagues. I was honored that he also chose me to become his deputy in our quite modest Institute. I was able to establish a small organic chemistry research group with some six to eight members, which also included my wife. Our laboratories were in rented space in an industrial research institute on the outskirts of Pest, but I certainly had more space than my original small "balcony laboratory" at the Technical University. Conditions, however, were still difficult. I never would have thought that this small research institute eventually would grow after my departure from Hungary into a large institution with hundreds of researchers, which is what I found when, after more than 25 years, I visited my native Hungary for the first time. In the post-Communist area, however, university research is again encouraged and the oversized research institutes of the Academy of Sciences are slowly being cut back to a more reasonable size.

We were very much isolated in Hungary from the scientific world, which is the worst thing that can happen to scientists. Journals and books were difficult to obtain, and even then only with great delays. I was able to attend only one scientific meeting in the West, the 1955 meeting of the International Union of Pure and Applied Chemistry

(IUPAC) meeting in Zurich, where I gave a lecture about some of our research. It was a unique experience, and I also met for the first time many chemists whose work I knew only from the literature. Little did I foresee that a year later my life was going to change drastically and such contacts would regularly become possible.

Contacts even with colleagues in the Soviet Union and other Eastern countries were difficult. In 1955–1956 I gave some lectures in what was then East Germany, including in East Berlin at the Humboldt University, and was able to see Emil Fischer's old laboratories. I also took the opportunity to cross into West Berlin (this was before the wall had gone up) to visit some of the organic chemists (Weygand, Bohlmann). At an East German scientific meeting I met some fine Czech colleagues, including Lukes [whose former graduate student was Vlado Prelog (Nobel Prize 1975)] and Wichterle (the later inventor of soft contact lenses and a very brave man who had a significant part in the Czech Velvet Revolution in the late 1960s). My closest contact was with Costin Nenitzescu from Bucharest, Rumania. He visited Budapest on several occasions. We met again in the late 1960s in Cleveland and kept in close touch until his untimely death. Interestingly, even contacts with the Soviet Union were weak. Remarkably, information about Russian research at the time was easiest to obtain from the *Chemical Abstracts* published by the American Chemical Society. In any case, at this East German meeting I met Reutov and eventually through him Nesmeyanov, who not only was a leading chemist but also a great power in Soviet science, being the president of the Academy of Sciences, which had a formidable string of large research institutes. I visited the Soviet Union only once, accompanying Geza Schay, the director of our research institute. Because Nesmeyanov was interested in some of my work in areas where he himself had worked, when we were in Moscow I was taken to see him in his Institute of Organoelement Compounds, a large and impressive place with many hundreds of researchers. We had a good discussion in his office. We spoke French and German (I don't speak Russian), but, surprisingly, when somebody came in to join us, he changed to Russian and asked for an interpreter. Even he seemed to be concerned at the time not to be reported for such "anti-Soviet" behavior. In our private conversation, Nesmeyanov told me about the recent condemnation of Pauling's resonance theory. He mentioned that he had not signed the official report orchestrated by the political powers, claiming illness so as not to participate in the process. Even in retrospect, it showed a most discouraging picture of the political influence at the time on Soviet science.

Other memories I have about my visit involve Kitaigorodsky, an excellent X-ray crystallographer. He opened the door to a large room filled with tables at which sat row upon row of girls cranking mechanical calculators. With a sad smile, he said that this was his computer, but he had three shifts of girls working at the calculators. The situation was different at Semenov's Institute of Chemical Physics (he received the Nobel Prize, shared with Norrish, in the following year, 1956). He told us in some detail about his state-of-the-art work on shock waves and high-energy propulsion systems. I realized only after Sputnik was launched in 1957 that his work must have had some connection with it.

Isolation clearly was a most depressing aspect of pursuing science in Communist-dominated Hungary. To overcome it, sometimes one even tried to attempt unusual paths. In 1955, I read in a scientific journal an announcement about the Dutch Van't Hoff Fellowship, which offered to the winner the possibility to visit Holland to meet with Dutch scientists. Although I had no permission (it may seem strange, but at the time we were not allowed even such contacts with the West), I applied for it, not expecting to hear anything further about it. To my great surprise, however, I was chosen for the fellowship in early 1956, but I was not given the opportunity at the time to visit Holland. However, in 1957, by which time I lived in Canada, I was able to lecture at an IUPAC conference in Paris and, on my way, visit Holland. I greatly enjoyed meeting Dutch colleagues, particularly Professors Wibaut and Sixma, and enjoyed their gracious hospitality.

I spent two very productive years in the Chemical Research Institute until the fall of 1956, making the best use of existing research possibilities. In October 1956, Hungary revolted against Soviet rule, but the uprising was soon put down by drastic measures after intense fighting with overwhelming Soviet forces and much loss of life. Budapest was again devastated, and the future looked bleak. Our research institute was not itself damaged during the fighting, but our spirit was broken. During the short-lived days of freedom, a small revolutionary committee was formed in our Institute to which I was elected. There was nearly unanimous support for the spontaneous revolt of 1956 in Budapest, but it could not on its own prevent the oppression and terror that returned with the Soviet forces. Consequently, in November and early December of 1956 some 200,000 Hungarians, mostly of the younger generation, fled their homeland. With my family (our older son, George, was born in 1954) and most of my research group, who also decided that this was the only path to follow, I joined the torrent of refugees seeking a new life in the West.

· 6 ·

Move to North America:
Industrial Experience While Pursuing the Elusive Cations of Carbon

After fleeing Hungary with most of my research group in the exodus of November of 1956, in early December, my wife, our 2-year-old son, and I eventually got to London, where an aunt of my wife lived. We were warmly welcomed. During our stay in London I was able to meet for the first time some of the chemists whose work I knew from the literature and admired. I found them most gracious and helpful. In particular, Christopher Ingold and Alexander Todd (Nobel Prize in chemistry 1957) extended their efforts on behalf of a young, little-known Hungarian refugee chemist in a way that I will never forget and for which I will always be grateful.

While in London I visited Ingold at University College and also met some of his colleagues, including Ed Hughes, Ron Nyholm, and Ron Gillespie (with whom I later renewed contact in Canada). I was invited in early January 1957 to give a seminar at Cambridge. This was the first time I ever lectured in English. My talk raised some comments (which I understand were made before in connection with another Hungarian émigré's similar experience) on "how interesting it is that this strange Hungarian language has some words which resemble English." Anyhow, I survived my "baptism," but, like many Hungarian-born scientists, I retain to this day an unmistakable accent (although some believe that this may be an asset). I remember about my visit that I was picked up at the railroad station by my host Bob Haszeldine (whose work I knew through my work in fluorine chemistry), who was driving what turned out to be a pricey Bentley. He was a lecturer at the time and was helping Todd with the design of the new Lensfield Road chemistry building. (Later he moved to Manchester and had a major part in building the Manchester Institute of Technology.) In late afternoon, I was dropped off at the same station by Todd in his small Morris. I did not yet understand the significance some in Western society attach to the kind of car they drive and its assumed social implication; Todd certainly was not one of them. Another memory of my visit was a conversation in which Al Katritzky, another young member of the faculty who was preparing to join the planned new "red brick" University of East Anglia, was telling Todd about his plans for the new chemistry department. They included even a number of NMR spectrometers, quite extravagant at the time. Todd listened politely and then remarked: "if your plans work out properly, in a few centuries it indeed may become a good university."

A heartwarming experience for my wife and me was meeting Ms. Esther Simpson, who was running the Academic Assistance Counsel (AAC) and had a modest office off Piccadilly next to Burlington House. She wrote many letters on my behalf trying to help with my search for a job and our resettling. She also extended constant kindness and encouragement to us. I learned years later that Leo Szilard was essential in helping to organize AAC as a clearinghouse to match refugees of the Hitler era with placement possibilities. Ms. Simpson was the essential dynamo of the organization and helped many refugee scientists. My own situation was probably not different from that faced by many others over the years who learned to appreciate her warm humanity.

We did not intend to settle in England and were looking forward to moving to Canada; my mother-in-law lived in Montreal, having remarried there after the war. Todd, having learned about our plans to move to Canada, strongly recommended that I go to Saskatoon, Saskatchewan, where he had a nephew, a professor in the medical school, whom he visited on occasion. He felt the university there was a pleasant, promising place. When I looked it up, however, it looked very much to be in the middle of the prairie, wide open as far as the North Pole. It turned out that Todd's visits were always during the summer.

Having learned more about the long, harsh winters of Saskatoon, we were discouraged from following his advice. (I learned later that Gerhardt Herzberg spent years in Saskatoon before moving to Ottawa, but he was probably a sturdier person.) In March 1957, we traveled by air to Montreal, and I renewed my efforts of looking for a job to be able to support my family. I felt that perhaps in my native small Hungary I was already an established researcher with some recognized achievements, but in my search I learned fast that in the new world this meant little. Canada at the time (perhaps not much differently from the present) had few research opportunities. At Ingold and Todd's recommendation, Maurice Steacie, who was heading the National Research Council in Ottawa, offered me a postdoctoral fellowship after a visit there, but I needed a permanent job. While I was initially looking for an academic position, none came along, but a few industrial research possibilities opened up. Of these, that at Dow Chemical was the most interesting.

Dow Chemical, with its home base in Midland, Michigan, was establishing at the time a small exploratory research laboratory 100 miles across the Canadian border in Sarnia, Ontario, where its Canadian subsidiaries' major operations were located. I was offered the opportunity to join this new laboratory, and two of my Hungarian collaborators who also came to Canada, including Steve Kuhn, could also join me. We moved to Sarnia in May of 1957. As our moving expenses were paid for, with a feeling of extravagance we checked in the two cardboard boxes containing all our worldly possessions for the train trip and started out for our destination on the shore of Lake Huron, fifty miles north of Detroit. Our younger son, Ronald, was born in Sarnia in 1959, and we have fine memories of the seven years we spent there. It was a pleasant, small industrial city with a large concentration of petrochemical industries and refineries, but the residential areas were in the north of town, close to the lake and away from the industrial area. The surroundings were well suited to bringing up a young family. There was, however, no possibility for Judy to continue her career in research, because industry at the time would not employ a married couple. However, she rejoined our joint effort when in 1965 we moved to Cleveland and I returned to academic life.

The Olah family 1962

In the 1950s it was a general trend for major American chemical companies to establish European research laboratories. For example, Cyanamid in Geneva, Union Carbide in Brussels, and Monsanto in Zurich set up such laboratories with impressive staffs of fine chemists such as Hudson, Jorgensen, and Klopman at Cyanamid, Dahl, Schröder, and Viehe at Union Carbide, Zeiss at Monsanto, and others. Dow kept out of Europe for a long time (perhaps because of the lingering effect of its pre-World War II German industrial contacts, which resulted in congressional inquiries during the war). Instead, in 1955 it established an Exploratory Research Laboratory in Canada financed by and responsible directly to the parent company. This small laboratory (with a staff of perhaps 15–20) had a significant degree of freedom and close contact with Dow research in Midland, MI, just two hours drive across the border from Sarnia.

As a chemist, it was easy for me to fit into the new environment. Science is international and even has its own language in scientific English (as it is frequently called), i.e., English with a foreign accent. The initial language difficulty many immigrants face is certainly much easier to overcome in the technical and scientific fields.

What was more of a challenge was adapting to the ways of an American-style laboratory and its research practices. Young research chemists were paid salaries not significantly higher than those of technicians, who were, however, unionized. Work was consequently carried out mostly by Ph.D. chemists with few technical support personnel (a situation different from European practices of the time). We set up our laboratory and it was possible to restart active research soon after my arrival at Dow. I was grateful and still am for the opportunity and support given to me. I also look back with some envy and nostalgia to this period, when I spent most of my time working at the bench doing experimental work. This, however, still left the evenings and weekends for pursuing my own research interests, for reading, writing, and thinking about new problems and projects.

After renting for a short while, we bought a small house close to Lake Huron quite a distance from the industrial area. We furnished it with all the necessities of our new life, as we literally had nothing when we arrived. For two years we had no car, still believing at the time in such old-fashioned principles as not going into debt and paying cash for everything we purchased. This, of course, caused problems in getting around, including going to work, relying on spotty public bus service or getting rides from friends. When young, however, one notices such inconveniences much less.

I must also have made other impressions on my new colleagues with my work habits. In an industrial laboratory, located within a chemical plant complex, nobody thought of working extra-long hours and weekends, except for some "strange" scientists. However, what I did "after hours" was conceded to be research to pursue my own interests. This suited me well, and I enjoyed the opportunity and pursued it vigorously. In the eight years I spent with Dow I published some 100 scientific papers (on which I did at least part of the experimental work myself). I also obtained some 30 patents on various topics of industrial interest. Some of my other discoveries, for example, on the beginnings of superacidic systems and related chemistry, however, were not considered worth patenting, and many of my internal discovery memoranda were never developed into patent applications. On my own time I also worked on revising and updating the German version of my book, *Theoretical Organic Chemistry*, which was originally published in Hungary. It was finally published in 1960. I also completed editing (and in no small part writing) a four-volume comprehensive monograph, *Friedel Crafts and Related Reactions*, published in 1963–1965 by Wiley-Interscience. This was a major effort, as it comprised close to 4000 pages and more than 10,000 references. Whereas I was given considerable freedom to pursue my research interests, a part of my research was also directed to questions and problems related to the company's industrial interests, which of course were not published in the literature.

My years at Dow were productive and rewarding. It was during this period in the late 1950s that my breakthrough work on long-lived stable carbocations, which in a way I had already started back in Hungary, was carried out. Dow was and is a major user of Friedel-Crafts-type chemistry, including the manufacture of ethylbenzene for styrene production by the reaction of benzene with ethylene. This was assumed to involve cationic intermediates (i.e., carbocations), which were, however, never observed or studied. My work thus also had practical significance and in some way helped to improve the large-scale industrial process. I was treated generously by Dow and was promoted rapidly to Research Scientist, the highest research position without direct administrative responsibility. I also gained much practical experience in the real, industrial world of chemistry, which served me well through the rest of my career, while at the same time continuing my own more academically oriented interests.

During my time at Dow I was also able to establish personal contact with many researchers whose work I previously knew only from the literature. Although our laboratory was located in a small Canadian industrial city off the path academic visitors usually frequent, I succeeded in starting a seminar program. An increasing number of outstanding chemists came, perhaps as a personal favor, to visit our small industrial laboratory, including Christopher Ingold, Georg Wittig, and Rolf Huisgen. I much appreciated their kindness. Another rewarding aspect of my years with Dow was that I was given the chance to publish much of my work and to lecture at different international symposia and meetings, such as the IUPAC Congresses in Paris (1957), Munich (1959), and London (1963); the Cork (Ireland) Conference on Physical Organic Chemistry (1961); and various meetings and symposia of chemical societies. I also used these occasions to visit and give lectures at many universities. My personal contacts and my circle of colleagues, many of whom over the years became friends, grew.

In the summer of 1963, I learned that I had won the American Chemical Society Award in Petroleum Chemistry for my work on Friedel-Crafts chemistry. It was a most welcome recognition for some-one who only a few years earlier had fled his native country and started all over on a far-away continent. Although I have received numerous other awards and recognitions over the years, with the exception of the Nobel Prize, no other award touched me as much. I remember that my first ACS award carried with it a check for $5,000. My research director for some reason believed that a company employee was not

1960 with Georg Wittig

In my "sailing" outfit, complete with jacket and tie, with Phil Skell and Heini Zollinger. Cork, Ireland, July 1964

At London IUPAC congress 1963

With my father in Austria during a European visit in 1965

entitled to any external "income" and I should give this to the company. He also raised some questions concerning royalties of any books I was writing in my spare time. The matter was finally settled by the president of the company, who found that these questions had no merit. I was able to keep my award money and modest book royalties (which I believe in any case would not have made any impact on the bottom line of the company's earnings).

Concerning my research during my Dow years, as I discuss in Chapter 4, my search for cationic carbon intermediates started back in Hungary, while I was studying Friedel-Crafts-type reactions with acyl and subsequently alkyl fluorides catalyzed by boron trifluoride. In the course of these studies I observed (and, in some cases, isolated) intermediate complexes of either donor-acceptor or ionic nature.

$$RCOF + BF_3 \rightleftharpoons \overset{\delta+}{R}CO \overset{\delta-}{\underset{F}{\longrightarrow}} BF_3 \rightleftharpoons RCO^+ BF_4^-$$

$$RF + BF_3 \rightleftharpoons \overset{\delta+}{R}F \overset{\delta-}{\longrightarrow} BF_3 \rightleftharpoons R^+ BF_4^-$$

Many elements readily form ionic compounds such as table salt (Na^+Cl^-), in which the cationic sodium and anionic chlorine are held together by electrostatic, ionic bonding. Carbon, however, was long considered to lack the ability to form similar ionic compounds, except in very specific, highly stabilized systems, such as triphenylmethyl dyes.

The Chicago chemist Stieglitz (whose noted photographer brother was the husband of the painter Georgia O'Keefe) suggested in 1899 the possibility of ionic carbon compounds, but it aroused no interest. When I gave the Stieglitz memorial lecture in the 1980s at the University of Chicago and reminded the audience of Stieglitz's pioneering idea, there was also little appreciation of it. It was in 1901 that Norris and Kehrman, as well as Wentzel, independently discovered that colorless triphenylmethyl alcohol gave deep yellow solutions in concentrated sulfuric acid. Triphenylmethyl chloride similarly formed orange complexes with aluminum and tin chlorides.

Von Baeyer (Nobel Prize, 1905) should be credited for having recognized in 1902 the saltlike character of the compounds formed. He then suggested a correlation between the appearance of color and salt formation—the so-called "halochromy." Gomberg (who had just shortly before discovered the related stable triphenylmethyl radical), as well as Walden, contributed to the evolving understanding of the structure of related cationic dyes such as malachite green.

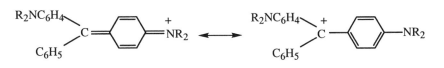

Whereas the existence of ionic triarylmethyl and related dyes was thus established around the turn of the twentieth century, the more general significance of carbocations in chemistry long went unrecognized. Triarylmethyl cations were considered an isolated curiosity of chemistry, not unlike Gomberg's triarylmethyl radicals. Not only were simple hydrocarbon cations believed to be unstable, even their fleeting existence was doubted.

One of the most original and significant ideas in organic chemistry was the suggestion by Hans Meerwein that carbocations (as we now call all the positive ions of carbon compounds) might be intermediates in the course of reactions that start from nonionic reactants and lead to nonionic covalent products.

In 1923, Meerwein, while studying the Wagner rearrangement of camphene hydrochloride to isobornyl chloride with van Emster, found that the rate of the reaction increased with the dielectric constant of the solvent. Furthermore, he found that certain Lewis acid chlorides such as $SbCl_5$, $SnCl_4$, $FeCl_3$, $AlCl_3$, and $SbCl_3$ (but not BCl_3 or $SiCl_4$) as well as dry HCl, which promote the ionization of triphenylmethyl chloride by formation of carbocationic complexes, also considerably accelerated the rearrangement of camphene hydrochloride to isobornyl chloride. Meerwein concluded that the isomerization actually does not proceed by way of migration of the chlorine atom but by a rearrangement of a cationic intermediate. Hence, the modern concept of carbocationic intermediates was born. Meerwein's views were, however, greeted with much skepticism by his peers in Germany, discouraging him from following up on these studies (see Chapter 9).

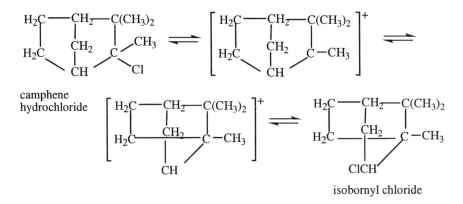

camphene hydrochloride

isobornyl chloride

Ingold, Hughes, and their collaborators in England, starting in the late 1920s, carried out detailed kinetic and stereochemical investigations on what became known as nucleophilic substitution at saturated carbon and polar elimination reactions. Their work relating to unimolecular nucleophilic substitution and elimination, called S_N1 and E1 reactions, in which formation of carbocations is the slow rate-determining step, laid the foundation for the role of electron-deficient carbocationic intermediates in organic reactions.

Frank Whitmore in the United States in the 1930s in a series of papers, generalized these concepts to include many other organic reactions. Carbocations, however, were generally considered to be unstable and transient (short lived) because they could not be directly observed. Many leading chemists, including Roger Adams, determinedly doubted their existence as real intermediates and strongly opposed even mentioning them. Whitmore, consequently, was never able, in any of his papers in the prestigious *Journal of the American Chemical Society*, to use the notation of ionic R_3C^+. The concept of carbocations, however, slowly grew to maturity through kinetic, stereochemical, and product studies of a wide variety of reactions. Leading investigators such as P. D. Bartlett, C. D. Nenitzescu, S. Winstein, D. J. Cram, M. J. S. Dewar, J. D. Roberts, P. v. R. Schleyer, and others contributed fundamentally to the development of modern carbocation chemistry. The role of carbocations as one of the basic concepts of modern chemistry has been well reviewed. With the advancement of mass spectrometry, the existence of gaseous carbocations was proven, but this could not give an indication of their structure or allow extrapolation to solution chemistry. Direct observation and study of stable, long-lived carbocations, such as alkyl cations in the condensed state, remained an elusive goal.

My work on long-lived (persistent) carbocations dates back to the late 1950s at Dow and resulted in the first direct observation of alkyl cations. Subsequently, a wide spectrum of carbocations as long-lived species was studied using antimony pentafluoride as an extremely strong Lewis acid and later using other highly acidic (superacidic) systems.

Until this time alkyl cations were considered only transient species. Their existence had been indirectly inferred from kinetic and stereochemical studies, but no reliable spectroscopic or other physical measurements of simple alkyl cations in solution or in the solid state were obtained.

It was not fully realized until my breakthrough using superacids (vide infra) that, to suppress the deprotonation of alkyl cations to olefins and the subsequent formation of complex mixtures by reactions of olefins with alkyl cations, such as alkylation, oligomerization, polymerization, and cyclization, acids much stronger than those known and used in the past were needed.

$$(CH_3)_3C^+ \rightleftharpoons H^+ + (CH_3)_2C{=}CH_2$$

Finding such acids (called "superacids") turned out to be the key to obtaining stable, long-lived alkyl cations and, in general, carbocations. If any deprotonation were still to take place, the formed alkyl cation (a strong Lewis acid) would immediately react with the formed olefin (a good π-base), leading to the mentioned complex reactions.

In Friedel-Crafts chemistry it was known that when pivaloyl chloride is reacted with aromatics in the presence of aluminum chloride, *tert*-butylated products are obtained in addition to the expected ketones. These were assumed to be formed by decarbonylation of the intermediate pivaloyl complex or cation. In the late 1950s I returned to my earlier investigations of Friedel-Crafts complexes and extended them by using IR and NMR spectroscopy. I studied isolable complexes of acyl fluoride with Lewis acid fluorides, including higher-valence Lewis acid fluorides such as SbF_5, AsF_5, and PF_5. In the course of these studies, it was not entirely unexpected that the generated $(CH_3)_3CCOF$-SbF_5 complex showed a substantial tendency toward decarbonylation. What was exciting, however, was that it was possible to follow this process by NMR spectroscopy and to observe what turned out to be the first stable, long-lived alkyl cation salt, namely, *tert*-butyl hexafluoroantimonate.

$$(CH_3)_3CCOF + SbF_5 \longrightarrow (CH_3)_3CCO^+ SbF_6^- \xrightarrow{-CO} (CH_3)_3C^+ SbF_6^-$$

This breakthrough was first reported in 1962 and was followed by further studies that led to methods for preparing varied long-lived alkyl cations in solution.

The idea that ionization of alkyl fluorides to stable alkyl cations could be possible with an excess of strong Lewis acid fluoride that also serves as solvent first came to me in the early 1950s while I was still working in Hungary and studying the boron trifluoride-catalyzed alkylation of aromatics with alkyl fluorides. In the course of these studies I attempted to isolate $RF:BF_3$ complexes. Realizing the difficulty of finding suitable solvents that would allow ionization but at the same would not react with developing, potentially highly reactive alkyl cations, I condensed alkyl fluorides with neat boron trifluoride at low temperatures. I had, however, no access to any modern spectrometers to study the complexes formed. I remember a visit at the time by Costin Nenitzescu (an outstanding but not always fully recognized Romanian chemist, who carried out much pioneering research on acid-catalyzed reactions). We commiserated on our lack of access to even an IR spectrometer. (Nenitzescu later recalled sending his cyclobutadiene-Ag$^+$ complex on the Orient Express to a colleague in Vienna for IR studies, but the complex decomposed en route.) All I could do at the time with my RF-BF$_3$ complexes was to measure their conductivity. The results showed that methyl fluoride and ethyl fluoride complexes gave low conductivity, whereas the isopropyl fluoride and *tert*-butyl fluoride complexes were highly conducting. The latter systems, however, also showed some polymerization (from deprotonation to the corresponding olefins). The conductivity data thus must have been to some degree affected by acid formation.

$$R\!-\!F + BF_3 \rightleftharpoons \overset{\delta+}{R}\!-\!\overset{\delta-}{F}\!\rightarrow\!BF_3 \rightleftharpoons R^+BF_4^-$$

During a prolonged, comprehensive study some years later at the Dow laboratory of numerous other Lewis acid halides, I finally hit on antimony pentafluoride. It turned out to be an extremely strong Lewis acid and, for the first time, enabled the ionization of alkyl fluorides to stable, long-lived alkyl cations. Neat SbF$_5$ solutions of alkyl fluorides are viscous, but diluted with liquid sulfur dioxide the solutions could be cooled and studied at $-78°C$. Subsequently, I also introduced even lower-nucleophilicity solvents such as SO$_2$ClF or SO$_2$F$_2$, which allowed studies at even lower temperatures. Following up the observation of the decarbonylation of the pivaloyl cation that gave the first spectral evidence for the tertiary butyl cation, *tert*-butyl fluoride was ionized in excess antimony pentafluoride. The solution of the *tert*-butyl cation turned out to be remarkably stable, allowing chemical and spectroscopic studies alike.

In the late 1950s the research director of our laboratory was not yet convinced of the usefulness of NMR spectroscopy. Consequently, we had no such instrumentation of our own. Fortunately, the Dow laboratories in Midland just 100 miles away had excellent facilities run by E. B. Baker, a pioneer of NMR spectroscopy, who offered his help. To probe whether our SbF_5 solution of alkyl fluorides indeed contained alkyl cations, we routinely drove in the early morning to Midland with our samples and watched Ned Baker obtain their NMR spectra. *tert*-Butyl fluoride itself showed a characteristic doublet in its ^1H NMR spectrum due to the fluorine-hydrogen coupling ($J_{H.F} = 20$ Hz). In SbF_5 solution, the doublet disappeared and the methyl protons became significantly deshielded from about δ 1.5 to δ 4.3. This was very encouraging but not conclusive proof of the presence of the *tert*-butyl cation. If one assumes that with SbF_5 *tert*-butyl fluoride forms only a polarized donor-acceptor complex, which undergoes fast fluorine exchange (on the NMR time scale), the fluorine-hydrogen coupling would be "washed out," while a significant deshielding of the methyl protons would still be expected. The differentiation of a rapidly exchanging polarized donor-acceptor complex from the long-sought-after ionic $t\text{-}C_4H_9^+ SbF_6^-$ thus became a major challenge.

$$(CH_3)_3C\text{—}F + SbF_5 \rightleftharpoons (CH_3)_3C \overset{F}{\underset{F}{\diagup}} \overset{\overset{F}{|}}{\underset{\underset{F}{|}}{Sb}} \overset{F}{\diagdown} \quad \text{or} \quad (CH_3)_3C^+ \ SbF_6^-$$

Ned Baker, himself a physicist, showed great interest in our chemical problem. To solve it, he devised a means to obtain the carbon-13 NMR spectra of our dilute solutions, an extremely difficult task at the time before the advent of Fourier transform NMR techniques. Labeling with carbon-13 was possible at the time only to about a 50% level (from $Ba^{13}CO_3$). When we prepared 50% ^{13}C-labeled *tert*-butyl fluoride, we could, however, obtain at best only a 5% solution in SbF_5. Thus the ^{13}C content in the solution was tenfold diluted. However, Baker, undaunted, devised what became known as the INDOR (internuclear double resonance) method. Using the high sensitivity of the proton signal, he was able with the double-resonance technique to observe the ^{13}C shifts of our dilute solutions—a remarkable achievement around 1960. The carbon-13 shift of the tertiary carbon atom in $(CH_3)_3CF$-SbF_5 of $\delta^{13}C$ 335.2 turned out to be more than 300 ppm deshielded from that of the covalent starting material. Such very large chemical deshielding (the most deshielded ^{13}C signal at the time) could not be reconciled with a donor-acceptor complex. It indicated rehybridization from sp^3 to sp^2 and at the same time showed the effect of significant positive charge on the carbocationic carbon center.

Besides the *tert*-butyl cation, we also succeeded in preparing and studying the related isopropyl and the *tert*-amyl cations. The isopropyl cation was of particular relevance.

$$(CH_3)_2CHF + SbF_5 \longrightarrow (CH_3)_2CH^+ \, SbF_6^-$$

$$(CH_3)_2CFCH_2CH_3 + SbF_5 \longrightarrow (CH_3)_2\overset{+}{C}CH_2CH_3 \, SbF_6^-$$

Whereas in the *tert*-butyl cation the methyl protons are attached to carbons that are only adjacent to the carbocationic center, in the isopropyl cation a proton is directly attached to the center. When we obtained the proton NMR spectrum of the i-C_3H_7F-SbF_5 system, the CH proton showed up as an extremely deshielded septet at 13.0 ppm, ruling out a polarized donor-acceptor complex and indicating the formation of the $(CH_3)_2CH^+$ ion. The ^{13}C NMR spectrum was also conclusive, showing a very highly deshielded (by $\Delta\delta > 300$) ^+C atom ($\delta^{13}C$ 320.6). The spectrum of the *tert*-amyl cation showed an additional interesting feature due to the strong long-range H-H coupling of the methyl protons adjacent to the carbocationic center with the methylene protons. If only the donor-acceptor complex were involved, such long-range coupling through an sp^3 carbon would be small (1–2 Hz). Instead, the observed significant coupling (J_{H-H} = 10 Hz) indicated that the species studied indeed had an sp^2 center through which the long-range H-H coupling became effective. Figure 6.1 reproduces the ^1H NMR spectra of the *tert*-butyl, *tert*-amyl, and isopropyl cations. These original spectra are framed and hang in my office as a memento, as are the ESCA spectra of the *tert*-butyl and of the norbornyl cation (vide infra).

Our studies also included IR spectroscopic investigation of the observed ions (Fig. 6.2). John Evans, who was at the time a spectroscopist at the Midland Dow laboratories, offered his cooperation and was able to obtain and analyze the vibrational spectra of our alkyl cations. It is rewarding that, some 30 years later, FT-IR spectra obtained by Denis Sunko and his colleagues in Zagreb with low-temperature matrix-deposition techniques and Schleyer's calculations of the spectra showed good agreement with our early work, considering that our work was

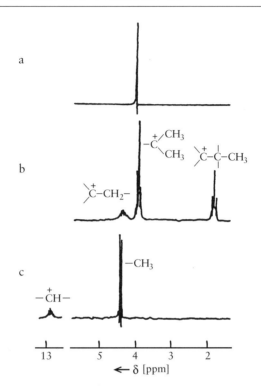

Figure 6.1. H NMR spectra of (*a*) the *tert*-butyl cation [trimethylcarbenium ion, $(CH_3)_3C^+$]; (*b*) the *tert*-amyl cation [dimethylethylcarbenium ion, $(CH_3)_3C^+$-C_2H_5]; and (*c*) the isopropyl cation [dimethylcarbenium ion, $(CH_3)_2C^+H$] (60 MHz, in SbF_5:SO_2ClF solution, $-60°C$).

carried out in neat SbF_5 at room temperature long before the advent of the Fourier transform methods. Subsequently, in 1968–1970 with Jack DeMember and Auguste Commeyras in Cleveland, we were able to carry out more detailed IR and laser Raman spectroscopic studies of alkyl cations. Comparison of the data of unlabeled and deuterated-*tert*-butyl cations with those of isoelectronic trimethylboron proved the planar structure of the carbocation.

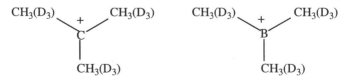

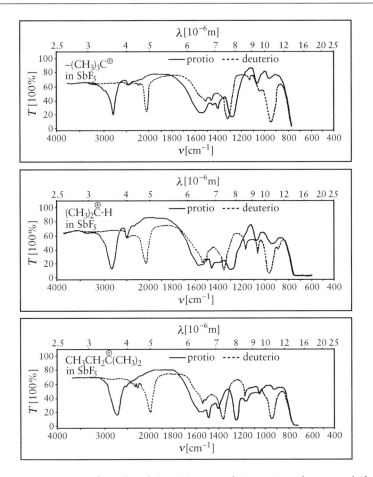

Figure 6.2. IR spectra of *tert*-butyl (*top*), isopropyl (*center*), and *tert*-amyl (*bottom*) cations. *T* = transmission.

This was also an early example of the realization that for nearly all carbocations there exists a neutral isoelectronic isostructural boron analogue, which later proved itself so useful in the hands of my colleagues R. E. Williams, G. K. S. Prakash, and L. Field.

In the summer of 1962, I was able for the first time to present my work in public at the Brookhaven Organic Reaction Mechanism Conference, and subsequently in a number of other presentations and publications. I had convincing evidence to substantiate that, after a long and frequently frustrating search, stable alkyl cations had finally been obtained in superacidic solutions.

The chemistry of stable, long-lived (or persistent) carbocations, as they became known, thus began and its progress was fast and widespread. Publication of research done in an industrial laboratory is not always easy. I would therefore like to thank again the Dow Chemical Company for allowing me not only to carry out the work but eventually also to publish the results.

Many have contributed since to the study of long-lived carbocations. The field rapidly expanded and allowed successful study of practically any carbocationic system. My talented and hard-working former associates and students deserve the lion's share of credit for our work, as do the many researchers around the world who joined in and contributed so much to the development of the field. (Their work can be found in the recommended readings.) I would like, however, to mention particularly the pioneering work of D. M. Brouwer and H. Hogeveen, as well as their colleagues at the Shell Laboratories in Amsterdam in the 1960s and 1970s. They contributed fundamentally to the study of long-lived carbocations and related superacidic hydrocarbon chemistry. The first publication from the Shell laboratories on alkyl cations appeared in 1964, following closely my initial reports of 1962–1963.

In the spring of 1964 I transferred from Sarnia to Dow's Eastern Research Laboratories in Framingham, Massachusetts, outside Boston, established under the directorship of a friend, Fred McLafferty, and located initially in a converted old industrial building. When Christopher Ingold came to visit us there one day later in the year, he had difficulty in convincing the cab driver who drove him from his hotel in Boston to find the "noted research laboratory" in the dilapidated neighborhood the address indicated. The laboratory, however, soon moved to neighboring Wayland, into nice a campuslike setting. Fred built up an impressive laboratory in a short time, where it was possible to work under the most pleasant conditions. Harvard and MIT were just a half hour driving distance away and provided valuable contact with the academic community. Bill Lipscomb (Nobel Prize in chemistry,

1976) and Paul Bartlett from Harvard became consultants and visited regularly. The Harvard and MIT seminars provided extraordinary opportunities for participation. In 1965, however, Fred McLafferty decided to leave for academic life (first at Purdue and then Cornell). Shortly thereafter, I also felt it was time to move back to academia and followed his example.

·7·

Return to Academia—
The Cleveland Years:
Carbocations, Magic Acid, and
Superacid Chemistry

In Canada, following my escape from Hungary in 1956, I found no opportunity to continue my academic career. Around 1960, I had a brief visiting professorship at the University of Toronto, to which Alexander Todd had previously recommended me through Charles Best. Some interest was expressed in me, but nothing developed. Years later, quite unexpectedly, I received a letter from George Wright, a senior organic faculty member at Toronto, expressing his regret that he had strongly opposed the appointment of a young Hungarian refugee chemist with no proper credentials and school ties, who he believed did not have much to offer. Belatedly, he wanted to tell me that he had been wrong. I appreciated his letter, but, as with many other things in life, one never knows what will eventually turn out to be beneficial for the future. The failure to obtain an academic appointment certainly worked out this way in my case. I still have friends and even some family in Canada (sons of my late cousin live in Toronto). Our younger son, Ron, was born there in 1959, and we have only pleasant memories of our Canadian years. After I received my Nobel Prize I was evidently rediscovered in Canada; for example, the Royal Society of Canada elected me as a Foreign Fellow, which I appreciate.

During my subsequent stay at the Dow Eastern Research Laboratory in Boston, I regularly attended and participated in the weekly Bartlett-Westheimer seminars at Harvard. I must have made some impression, because early in 1965 Paul Bartlett asked me whether I would be interested in going back to academia. He was regularly consulting in the Cleveland-Akron area and had been asked to recommend someone as professor and chairman for the chemistry department at Western Reserve University in Cleveland. This was probably not the kind of appointment most of his former students or acquaintances were interested in. I did appreciate his help, however, and was not discouraged by a move to a small Midwestern university. The Western Reserve, as I learned, was the vast reserve land that Connecticut was given in colonial times in case its crops were burned down by the Indians or ruined by some natural calamity. It was considered at the time to be on the edge of civilization but later became part of the heartland of America.

In the summer of 1965, we moved to Cleveland and I took on my duties including the chairmanship of the chemistry department. Upon my arrival in Cleveland I got much collegial help and advice from Professor Frank Hovorka, who had just retired as chair of the chemistry department, from Ernest Yeager, a leading electrochemist and faculty colleague, who acted as interim chair, and from my other new colleagues. I hope that I lived up to their expectations. Throughout my life I have always put great importance on loyalty, which should work both ways. It has always guided me in my relationships with colleagues and students alike, and my experience in Cleveland reinforced its importance for me.

During my chairmanship in Cleveland I was able to attract some fine colleagues to join our efforts to build a much-improved chemistry department. Among others, Miklos Bodanszky, an old friend from Budapest and an outstanding peptide chemist, came from Squibb in Princeton. With his wife Agnes (an established researcher in her own right), he soon established an excellent and most active research group in peptide synthesis. Gilles Klopman, whom I first met during a visit to the Cyanamid research laboratory in Geneva, joined us via a stay with Michael Dewar in Austin, Texas. He started one of the pioneering efforts in computational chemistry. Rob Dunbar came from John Baldeschwieler's group at Stanford and continued his work in ion cyclotron resonance spectroscopy. Muttaya Sundaralingam came and established a fine laboratory in structural X-ray crystallography. We also attracted Bill Stephenson from Stanford, a dynamic physical organic chemist, who after a very productive time at Case Western Reserve, joined us in Los Angeles and helped to build the Loker Institute, before deciding to follow other directions in his career. George Mateescu, who originally came to do research with me from Costin Nenitzescu's Institute in Bucharest, decided to stay in America and eventually also joined the faculty. In addition to his own research, he built an outstanding instrument facility centered around magnetic resonance spectroscopy, mass spectrometry, and X-ray photoelectron spectroscopy, at the same time fostering close cooperation with industry.

One achievement of my Cleveland years of which I am particularly proud was to have succeeded in combining the chemistry departments of Western Reserve University and the neighboring Case Institute of Technology. The two departments occupied practically adjacent buildings separated only by a parking lot. I had colleagues and good friends at Case, including John Fackler and Jay Kochi. It became obvious in the year after my arrival that it would make sense to join the two departments into a single, stronger department. We achieved this by 1967 with surprisingly little friction, and I was asked to stay on for a while as the chair of the joint department. I had never had administrative ambitions, and I was anxious to give up administrative responsibilities. In 1969 I rotated out of the chairmanship, and was named the C. F. Maybery Distinguished Professor of Research. John Fackler took over as chairman, and he did an outstanding job. I continued my research uninterrupted and never gave up my teaching; neither effort suffered during this period, which in fact was probably one of my most productive research periods. The merging of the two chemistry departments was so successful that it prompted the merger of Western Reserve University with Case Institute of Technology in 1970, forming Case Western Reserve University, which continues as a successful university.

I may have some talent for coping with administrative duties and dealing with people, but it never affected me in the ways I have sometimes observed it did others. When in my early years in Hungary I was helping to start a small research institute of the Academy of Sciences,

I considered my role mainly to try to help my colleagues while jealously safeguarding most of my time for research and scholarship. I proceeded in the same way in Cleveland and later in Los Angeles, when starting the Hydrocarbon Research Institute at USC. I have been fortunate never to have been bitten by the bug that makes many people feel important by exercising administrative "power."

During my time as department chair in Cleveland I discovered that my colleagues mainly came to see me with their problems. I soon realized, however, that although good things must surely have also happened to them, they usually felt it not worthwhile to mention these. In any case, I was able to establish a good rapport with my colleagues. They understood that I was trying to do my best to help to solve their problems within the limits of what I was able to do. This put us on the "same side." They even started to share with me some of their successes as well as their problems. My efforts were not always successful, but at least they knew that I always tried my best.

Regrettably, such close relationships are frequently missing in the academic setting. The ones wielding administrative "power" frequently feel obliged to automatically say "no" to whatever request is made, instead of trying to find a solution. Assumed "power" can have a strange effect on people. In my career, I have dealt with various research directors and vice presidents in industry and with department chairs, deans, provosts, and even presidents in academia, and I have learned that it is always dangerous to generalize. Some were outstanding people doing a fine job. Others, however, as a friend once characterized it, were just "swelling" in their assumed importance instead of "growing" in their jobs. There is unfortunately much infighting even in academia, perhaps—as somebody once said—because the stakes are so small. In any case, you learn much about human nature over the years, and to live with it. In academic life, leadership must be by example. I am convinced that nothing is as effective as a collegial spirit of cooperation. It always was and still is a privilege for me to try to help my colleagues and students career in any way I can. I also believe that some of the best academic administrators are those who are reluctant to give up their scholarship and teaching and are not only able to return full time to it but actively look forward to it (and really mean it). On the other hand, I also learned that many fine people make a deliberate choice to pursue an administrative career and are performing an essential role in leading our academic institutions and universities.

Another truth I learned about coping with administrative duties was that you can delegate much to capable and willing colleagues as well as to a small but capable staff (which you should not let grow excessively). In my experience the de facto time needed for making administrative decisions myself has never interfered with my regular faculty and scholarly work. This is probably the case on all levels of academia. Of course, there is also the temptation for many to spend time in never-ending meetings filled with much self-adulation. I believe that most of this is really quite unproductive and unnecessary. I always tried to be a "good citizen," never refusing committee assignments. By freely speaking my mind, however, even at the first meeting, I discovered a nearly foolproof pattern. Those running the committees almost never asked me back again and were glad to excuse me from their established "circles." I don't know whether my own experience is relevant to the general experience, but it certainly worked well for me. All this helped to preserve my ability to stay a teacher and researcher and kept me away from much other involvement. I remained "one of us" instead of being looked upon as "one of them."

Although I may be too optimistic, I hope that our universities will regain their original goal and spirit and not succumb further to bureaucratic pressures. Universities are there for the sake of the students and the faculty who teach them. It is also their mandate to broaden our knowledge and understanding through scholarly work. Administrators at all levels should always remember that they are there to facilitate the academic learning and research process. They should try to find solutions in a collegial spirit and not to "rule" in a bureaucratic way. At the same time, they must also be capable of being decisive and making decisions even when they are not popular at the time. Am I serious about all this? Of course I am, and so are most of my colleagues whose efforts in teaching and research make our profession an honorable and essential one in an increasingly materialistic world.

The small research group that followed me to Cleveland included two postdoctoral fellows who worked with me (the first ever at Dow), Chris Cupas and Chuck Pittman, as well as Mel Comisarow, an extremely bright, talented young technician who had been with me since my Sarnia days. I "rescued" him at the time when our research director declared him "difficult to get along with." Mel, a Canadian from Northern Alberta, became my first graduate student in Cleveland and went on to a successful academic career after doing his postdoctoral work with John Baldeschwieler at Stanford. At the University of British Columbia in Vancouver he co-developed with Jim Marshall Fourier transform ion cyclotron resonance spectrometry (FT-ICR), a powerful and most useful analytical method.

When I moved to Cleveland, Dow donated to me my by then quite extensive collection of laboratory samples and chemicals, as well as other specialized equipment. We packed them into a rent-a-truck and Chris Cupas, a native Bostonian, volunteered to drive it to Cleveland. Things turned out somewhat more eventful than expected, however. Chris jammed the truck into an underpass of the Memorial Drive on the Cambridge side of the Charles River, disregarding the height signs and causing a major traffic jam. Fortunately, Chris was "well connected" with the Boston Police Department, and after being extricated he eventually made it to Cleveland safely with our chemicals and equipment.

The transition back to academic life was surprisingly easy for me. My group grew and rapidly reached 15–20, about equally divided between graduate students and postdoctoral fellows. I attracted interested and enthusiastic graduate students as well as fine postdoctoral fellows. My group stayed at this level, with some fluctuation, over my subsequent academic career (now that I am not taking on graduate students anymore, it is becoming much smaller and it is composed only of postdoctoral fellows). I have always felt that to be an effective research advisor and mentor I must have close and regular personal contact with all my young associates. For me, it would have been impossible to do this with a larger group. Besides the pleasure of teaching again and advising graduate students, I started to receive a steady flow of postdoctoral fellows and visiting scientists from all corners of the world. Dick Chambers from Durham was one of my earliest visitors, spending a year with us. We became good friends and still maintain contact and see each other from time to time.

My postdoctoral fellows in Cleveland over the years were a truly international group. Major universities rightly attract their students and researchers to a significant degree by their reputation. Nobody would question the value to someone's resume of having studied or researched at Harvard, Stanford, Berkeley, or Caltech. Going to a little-known Midwestern university in Cleveland, however, carried no such benefits. Therefore I particularly appreciated the personal interest and devotion of those who took their chances in joining my group. Some of my American postdoctoral fellows in Cleveland included Marty Bollinger, Jack DeMember, Earl Melby, Eli Namanworth, Dany O'Brien, Tom Kiovsky, Charles Juell, Richard Schlosberg, Paul Clifford, Louise Riemenschneider, and David Forsyth. Others of my postdoctoral fellows came from many other countries of the world to my laboratory to stay for a year or two of research. Among others, from Great Britain came Ken Dunne, Dave Parker, and Tony White; from Germany, Joe Lukas and Herbert Mayr; from France, Jean Sommer, August Commeyras, Alain Germain, and Jean-Marc Denis; from Switzerland, Paul Kreinenbuhl and Michael Calin; from Austria, Peter Schilling; from Italy, Giuseppe Messina and Franco Pelizza; from Spain, Greg Asensio; from Japan, Masashi Tashiro, Shiro Kobayashi, Masatomo Nojima, Jun Nishimura, Yorinobu Yamada, Norihiko Yoneda, Iwao Hashimoto, Tohru Sakakibara, and Toshiyuki Oyama; from Australia, Phil Westerman, David Kelly, Bob Spear, and Subhash Narang; from Israel, Yuval Halpern, Josef Kaspi, and David Meidar, from Taiwan, Johnny Lin; from Hong Kong, Tse-lok Ho; from Romania, George Mateescu; and from Hungary, George Sipos. These associates, together with my graduate students, formed a fine research team.

The relationship of a professor and his or her graduate students is a very special one. One of the most enjoyable aspects of my life was being able to guide and help the development of my graduate students and seeing them succeed. My students included, besides Mel Comisarow (who followed me from Dow in Canada through Boston to Cleveland, where he became my first graduate student), such talented and highly motivated young people as Mark Bruce, Dan Donovan, James

Grant, Balaram Gupta, Alice Ku, Gao Liang, Henry Lin, Christine Lui, Chuck McFarland, Ripu Malhotra, Y. K. Mo, Dick Porter, Surya Prakash (who later in Los Angeles became my close colleague and a wonderful friend, see Chapter 8), George Salem, Jacob Shen, John Staral, Jim Svoboda, Paul Szilagyi, and John Welch.

It is frequently said that the scientific career of professors is made by their students and associates who de facto carry out their joint research. I am no exception. However, I first had to create my own little research "enclave" with an atmosphere and spirit conducive to carrying out our work. I also feel that I was able to motivate my students, to bring out from them talents and efforts that sometimes surprised even them. It was a most rewarding experience to see that most of my students, when they became interested and motivated, achieved much and turned themselves into excellent, productive, and increasingly independent researchers.

The future successful careers of my graduate students and postdoctoral fellows were pursued to a larger degree in industry than in academia. Our American academic institutions preferred (and still do) pedigrees from leading universities, whereas industry looked primarily for capable and adaptable chemists with the ability to fit in and achieve in the practical world. I know that my own departments at Case Western and later at USC would rarely consider a new faculty member coming from a "modest" background. It may not only be snobbery but also a conviction that good people must have been associated with the leading and best universities. It is, however, my belief (as someone who never benefited from association with a leading university) that you cannot always judge someone based only on "provenance." Individual ability and drive, as well as personality and stamina to stay the course, particularly when lacking access to outstanding students or facilities, also counts much. It certainly, however, must be a great advantage to be associated with leading institutions, which rather automatically provide the most favorable conditions and also the reputation that can facilitate someone's career. On balance, there is justification for this preference for a "pedigree," and our leading universities and their faculties certainly deserve their standing. If I were a young person starting out today and had the choice to join one of the leading chemistry departments (as a student, postdoc, or a young faculty member), I probably would do so myself. I do not regret, however, having gone the harder way, although the decision was made for me mostly by circumstances beyond my control. I always tried to make the best of my available circumstances and probably benefited from it.

One of the pleasures of being a university professor is the privilege of teaching. I always enjoyed it, and I still do. Over my career I have mainly taught different aspects of organic chemistry, both at the undergraduate and, later, increasingly at the graduate level. Because my own research over the years involved synthetic, mechanistic, and structural aspects of chemistry, teaching varied topics always came naturally to me. I hope that I was also able to convey to my students my personal experiences and an historical perspective of chemistry. I was fortunate to have actively participated in and known many of the key contributors of the chemistry of the second half of the twentieth century. This helped to establish a lively and somewhat personal atmosphere in my classes, which textbooks alone cannot provide. I also have always believed that it is essential to "reach" my audiences and to establish a close contact, whether lecturing to students or speaking at seminars, meetings, and congresses. I never use prepared texts—only occasional notes to remind me what topics to cover. This forces me to prepare mentally for my lectures. By necessity, it also ensures that my lectures always vary, even if given on the same topic. You challenge yourself to give your best, and at the same time you are also the best judge of how well you succeeded. I can see no reason why students should attend lectures by a professor who only recites a textbook or available lecture notes and does not convey new aspects to the students or challenge them to participate, something that a printed text alone or even video or internet presentation cannot do.

The real value of direct interaction between a lecturer and his audience is in the interest he can arouse by presenting the topic in a challenging and perhaps personal way as well as in the give-and-take interaction of direct exchange and participation. I have always tried in my lectures to give my students not only a presentation of facts and concepts, but, based on my own experience and involvement, to convey to them the fascination of exploring the intriguing world of chemistry, making it a joyful experience. I have also tried to inspire them to explore the topics discussed further on their own. This is the way students eventually grow into competent chemists, or at least ones who have a reasonable understanding and appreciation of chemistry. I also hope that my students were never bored by my lectures. However intriguing the new ways of electronic communication are, on-line courses cannot fully replace direct student-teacher interaction and its truly interactive give and take, which, I believe, will remain a fundamental part of the educational experience. I also believe that however useful electronically accessible literature search and retrieval systems are, printed books and periodicals in our libraries will remain essential for study and research alike. Reading about any topic inevitably leads to many other aspects to be considered that are not a priori obvious. Even just browsing sections of a library when we do not know exactly what we are looking for can be a stimulating and irreplaceable experience.

During my Cleveland years I also initiated an active seminar program with the participation of many leading chemists who came to visit us. This allowed us to create a lively and stimulating atmosphere in the department, which benefited students and faculty alike. In 1969 I organized the first of many subsequent international research symposia. The symposium was on carbocation chemistry and was attended by many of the major investigators in the field (Nenitzescu, Brown, Winstein, Dewar, Schleyer, Gillespie, Saunders, and others). When I moved to Los Angeles, these symposia became annual events. My own group in Cleveland held weekly meetings and research seminars, which remained a permanent feature over the years. On the basis of our mutual interest, Ned Arnett and John Pople (Nobel Prize in chemistry, 1998) and their research groups from Pittsburgh joined us at regular intervals, and we had joint monthly meetings alternating between Cleveland and Pittsburgh.

My research during the Cleveland years continued and extended the study of carbocations in varied superacidic systems as well as exploration of the broader chemistry of superacids, involving varied ionic systems and reagents. I had made the discovery of how to prepare and study long-lived cations of hydrocarbons while working for Dow in 1959–1960. After my return to academic life in Cleveland, a main aspect of my research was directed to the exploration of the chemistry of these persistent cations of carbon compounds (carbocations) and the fascinating new area of chemistry opened up by superacids. Particular interest was generated as a great variety of carbocations was found to be readily generated and studied in these enormously strong acid systems.

Over a decade of research, we were able to show that practically all conceivable carbocations could be prepared under what became known as "stable ion conditions" using various very strong acid systems (see discussion of superacids) and low nucleophilicity solvents (SO$_2$, SO$_2$ClF, SO$_2$F$_2$, etc.). A variety of precursors could be used under appropriate conditions, as shown, for example, in the preparation of the methylcyclopentyl cation.

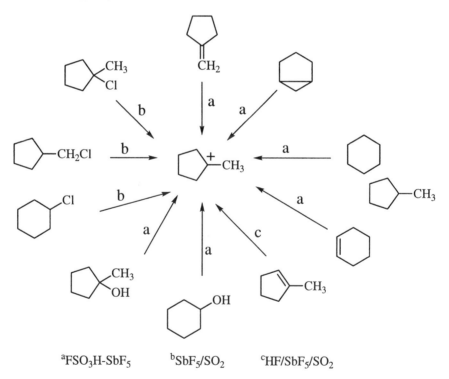

^{a}FSO$_3$H-SbF$_5$ ^{b}SbF$_5$/SO$_2$ ^{c}HF/SbF$_5$/SO$_2$

A wide variety of carbocations and carbodications, including those that are aromatically stabilized as well those as stabilized by heteroatoms, were reported in the nearly 200 publications on the topic during my Cleveland years.

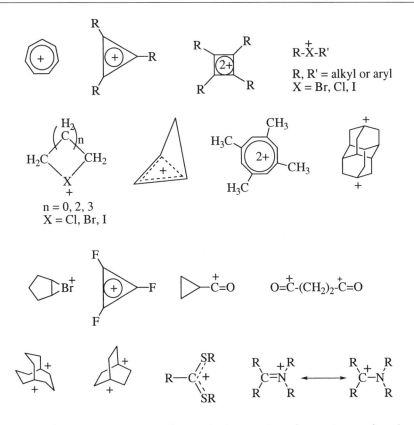

During this time I suggested (in 1972) naming the cations of carbon compounds "carbocations" (because the corresponding anions were named "carbanions"). To my surprise, the name stuck and was later officially adopted by the International Union of Pure and Applied Chemistry for general use.

Much effort was put into studying whether certain carbocations represent rapidly equilibrating or static (bridged, delocalized) systems (more about this in Chapter 9).

In the course of my studies, it became increasingly clear that a variety of highly acidic systems besides the originally used antimony penta-fluoride systems are capable of generating long-lived, stable carbocations. The work was thus extended to a variety of other superacids. Protic superacids such as FSO_3H (fluorosulfuric acid) and CF_3SO_3H (triflic acid) as well as conjugate acids such as $HF\text{-}SbF_5$, $FSO_3H\text{-}SbF_5$ (magic acid), $CF_3SO_3H\text{-}SbF_5$, and $CF_3SO_3H\text{-}B(O_3SCF_3)_3$ were extensively used. Superacids based on Lewis acid fluorides such as AsF_5,

Superacids in bottles. (Olah's Lifeblood)

TaF_5, and NbF_5 and other strong Lewis acids such as $B(O_3SCF_3)_3$ were also successfully introduced. The name "magic acid" for the FSO_3H-SbF_5 system was given by Joe Lukas, a German postdoctoral fellow working with me in Cleveland in the 1960s, who after a laboratory Christmas party put remainders of a candle into the acid. The candle dissolved, and the resulting solution gave a clear NMR spectrum of the *tert*-butyl cation. This observation understandably evoked much interest, and the acid used was named "magic." The name stuck in our laboratory. I think it was Ned Arnett who learned about it during one of his visits and subsequently introduced the name into the literature, where it became quite generally used. I helped a former graduate student of mine, Jim Svoboda, start a small company (Cationics) to make some of our superacidic systems and reagents commercially available, and he obtained trade name protection for Magic Acid. It has been marketed as such since that time.

I would like to credit especially the fundamental contributions of Ron Gillespie to strong acid (superacid) chemistry and also to recall his generous help while I was still working at the Dow Laboratories in Canada. I reestablished contact with him during this time. We first met in the winter of 1956 at University College in London, where he worked with Christopher Ingold. Subsequently, he moved to McMaster

University in Hamilton, Ontario. In the late 1950s, he had one of the early NMR spectrometers, and in our study of SbF_5 containing highly acidic systems and carbocations we appreciated his allowing us to run some spectra on his instrument. His long-standing interest in fluorosulfuric acid and our studies of SbF_5-containing systems thus found common ground in studies of FSO_3H-SbF_5 systems.

Until the late 1950s chemists generally considered mineral acids, such as sulfuric, nitric, perchloric, and hydrofluoric acids, to be the strongest acid systems in existence. This has changed considerably as extremely strong acid systems—many billions or even trillions of times stronger than sulfuric acid—have been discovered.

The acidity of aqueous acids is generally expressed by their pH, which is a logarithmic scale of the hydrogen ion concentration (or, more precisely, of the hydrogen ion activity). pH can be measured by the potential of a hydrogen electrode in equilibrium with a dilute acid solution or by a series of colored indicators. In highly concentrated acid solutions or with strong nonaqueous acids the pH concept is no longer applicable, and acidity, for example, can be related to the degree of transformation of a base to its conjugate acid (keeping in mind that this will depend on the base itself). The widely used so-called Hammett acidity function Ho relates to the half protonation equilibrium of suitable weak bases. The Hammett acidity function is also a logarithmic scale on which 100 percent sulfuric acid has a value of Ho -11.9. The acidity of sulfuric acid can be increased by the addition of SO_3 (oleum). The Ho of anhydrous HF is -11.0 (however, when HF is completely anhydrous, its Ho is -15, but even a slight amount of water drops the acidity to -11, as shown by Gillespie).

Perchloric acid ($HClO_4$; Ho -13.0), fluorosulfuric acid (HSO_3F; Ho -15.1), and trifluoromethanesulfonic acid (CF_3SO_3H; Ho -14.1) are considered to be superacids, as is truly anhydrous hydrogen fluoride. Complexing with Lewis acidic metal fluorides of higher valence, such as antimony, tantalum, or niobium pentafluoride, greatly enhances the acidity of all these acids.

In the 1960s Gillespie suggested calling protic acids stronger than 100% sulfuric acid "superacids." This arbitrary but most useful definition is now generally accepted. It should be mentioned, however, that the name "superacid" goes back to J. B. Conant of Harvard, who used it in 1927 in a paper in the *Journal of the American Chemical Society* to denote acids such as perchloric acid, which he found stronger than conventional mineral acids and capable of protonating such weak bases as carbonyl compounds. My book on superacids, published in 1985 with Surya Prakash and Jean Sommer, was appropriately dedicated to the memory of Conant. Few of today's chemists are aware of his contributions to this field. Conant subsequently became the president of Harvard University and gave up chemistry, which may explain why he never followed up on his initial work on superacids. At the end of World War II, he became the first allied high commissioner of occupied Germany and helped to establish a democratic West Germany. Upon his return home he continued a distinguished career in education and public service.

In a generalized sense, acids are electron pair acceptors. They include both protic (Brønsted) acids and Lewis acids such as $AlCl_3$ and BF_3 that have an electron-deficient central metal atom. Consequently, there is a priori no difference between Brønsted (protic) and Lewis acids. In extending the concept of superacidity to Lewis acid halides, those stronger than anhydrous aluminum chloride (the most commonly used Friedel-Crafts acid) are considered super Lewis acids. These superacidic Lewis acids include such higher-valence fluorides as antimony, arsenic, tantalum, niobium, and bismuth pentafluorides. Superacidity encompasses both very strong Brønsted and Lewis acids and their conjugate acid systems.

Friedel-Crafts Lewis acid halides form with proton donors such as H_2O, HCl, and HF conjugate acids such as $H_2O \cdot BF_3$, $HCl\text{-}AlCl_3$, and $HF\text{-}BF_3$, which ionize to $H_3O^+BF_3OH^-$, $H_2Cl^+AlCl_4^-$, and $H_2F^+BF_4^-$, etc. These conjugate Friedel-Crafts acids have Ho values from about -14 to -16. Thus they are much stronger than the usual mineral acids. Even stronger superacid systems are $HSO_3F\text{-}SbF_5$ (magic acid), HF-SbF_5 (fluoroantimonic acid), and $CF_3SO_3H\text{-}B(O_3SCF_3)_3$ (triflatoboric acid). The acidity of anhydrous HF, HSO_3F, and CF_3SO_3H increases drastically upon addition of Lewis acid fluorides such as SbF_5, which form large complex fluoroanions facilitating dispersion of the negative charge.

$$2\,HF + 2\,SbF_5 \rightleftharpoons H_2F^+\,Sb_2F_{11}^- \quad \text{fluoroantimonic acid}$$

$$2\,HSO_3F + 2\,SbF_5 \rightleftharpoons H_2SO_3F^+\,Sb_2F_{10}(SO_3F)^- \quad \text{magic acid}$$

$$2\,CF_3SO_3H + B(O_3SCF_3)_3 \rightleftharpoons CF_3SO_3H_2^+\,B(O_3SCF_3)_4^- \quad \text{triflatoboric acid}$$

The acidity function of HSO_3F increases on addition of SbF_5 from -15.1 to -23.0, the acidity of 1:1 FSO_3H-SbF_5 (magic acid). Fluoroantimonic acid is even stronger; with 4 mole percent SbF_5, the Ho value for HF-SbF_5 is already -21.0, a thousand times stronger than the value for fluorosulfuric acid with the same SbF_5 concentration. The acidity of the 1:1 HF-SbF_5 system or those with even higher SbF_5 concentrations reaches Ho -28. Thus these superacidic systems can be 10^{16} times stronger than 100 percent sulfuric acid! (A trillion is 10^{12}.)

Related superacid systems in which SbF_5 is replaced by AsF_5, TaF_5, NbF_5, etc. are of somewhat lower acidity but are still extremely strong acids. So is HF-BF_3, a very useful superacid that will not cause oxidative side reactions. Ternary superacid systems including, for example, FSO_3H-HF or CF_3SO_3H-HF with Lewis acid fluorides are also known and used.

Acids are not limited to liquid (or gaseous) systems. Solid acids also play a significant role. Acidic oxides such as silica, silica-alumina, etc. are used extensively as solid acid catalysts. New solid acid systems that are stronger than those used conventionally are frequently called solid superacids.

As applications of liquid superacids gained importance, attention was directed to finding solid or supported superacids suitable as catalysts. There are considerable difficulties in achieving this goal. For example, BF_3 cannot be well supported on solids because its high volatility makes its desorption inevitable. SbF_5, TaF_5, and NbF_5 have much lower vapor pressures and are thus much more adaptable to being supported or attached to solids. Because of their high chemical reactivity, SbF_5, HF-SbF_5, HSO_3F-SbF_5, etc. are preferentially supported on fluoridated alumina or fluorinated graphite. Solid superacids based on TaF_5 or NbF_5 are more stable than those based on SbF_5 because of their higher resistance to reduction. Solid perfluorinated resinsulfonic acid catalysts, such as those based on the acid form of DuPont's Nafion ionomer membrane resin, and some higher perfluoroalkanesulfonic acids, such as perfluorodecanesulfonic acid, have gained use as solid superacid catalysts. Some of the widely used zeolite catalysts (based on aluminosilicates, phosphates, etc.), such as HZSM-5, are also recognized to possess high acidity.

Starting in the late 1950s, I was fortunate to have found superacidic antimony pentafluoride, magic acid, and other related systems and to be able to explore their remarkable chemistry. The strength of some of these acids can be up to trillions of times stronger than that of concentrated sulfuric acid. Such large numbers have little meaning in our everyday life, and it is even difficult to comprehend their magnitude. As a comparison, the U.S. national debt is about 6 trillion dollars (6 $\times 10^{12}$). Superacids are indeed extremely strong, considering that they are obtained in the condensed state, where the "naked" proton cannot exist. In the gas phase—for example, in a mass spectrometer at high vacuum—in contrast, the proton can be unencumbered, i.e., "naked." This could be estimated to add an additional 30–35 powers of tens to the imagined (or unimaginable) Ho acidity. In the condensed state, a proton lacking any electron will always attach itself to any electron donor. It is with this caveat that we use "H^+" to denote the proton when discussing its role in condensed-state chemistry.

The high acidity of superacids makes them extremely effective protonating agents and catalysts. They also can activate a wide variety of extremely weakly basic compounds (nucleophiles) that previously could not be considered reactive in any practical way. Superacids such as fluoroantimonic or magic acid are capable of protonating not only π-donor systems (aromatics, olefins, and acetylenes) but also what are called σ-donors, such as saturated hydrocarbons, including methane (CH_4), the simplest parent saturated hydrocarbon.

Protonated methane (CH_5^+) does not violate the octet rule of carbon. A bonding electron pair (responsible for covalent bonding between C and H atoms) is forced into sharing with the proton, resulting in 2 electron-3 center bonding (2e-3c) (see Chapter 10). Higher alkanes are protonated similarly.

Dihydrogen (H_2) is similarly protonated to H_3^+ by superacids, as was shown by studies using isotopic labeling. The structure of H_3^+ again involves 2e-3c bonding.

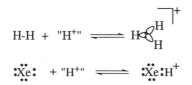

Nonbonded electron pair donors (*n*-donors) are expectedly readily protonated (or coordinated) with superacids. Remarkably, this includes even xenon, long considered an "inert" gas. The protonation of some π-, σ- and *n*-bases and their subsequent ionization to carbocations or onium ions is depicted as follows:

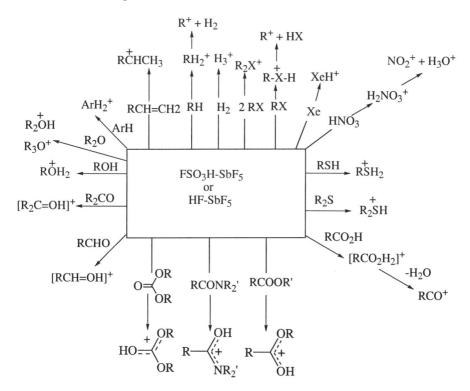

As expected, superacids were found to be extremely effective in bringing about protolytic transformations of hydrocarbons.

Isomerization (rearrangement) of hydrocarbons is of substantial practical importance. Straight-chain alkanes obtained from petroleum

oil generally have poorer combustion properties (expressed as low octane numbers of gasoline) than their branched isomers, hence the need
to convert them into the higher-octane branched isomers. Isomeriza-
tions are generally carried out under thermodynamically controlled
conditions and lead to equilibria. As a rule, these equilibria favor increased amounts of the higher-octane branched isomers at lower temperatures. Conventional acid-catalyzed isomerization of alkanes is carried out with various catalyst systems at temperatures of 150–200°C.
Superacid-catalyzed reactions can be carried out at much lower temperatures (even at or below room temperature), and thus they give
increased amounts of preferred branched isomers. This, together with
alkylation, is of particular importance in the manufacture of lead-free
high-octane gasoline. We have studied these processes extensively.

Alkylation combines lower-molecular-weight saturated and unsaturated hydrocarbons (alkanes and alkenes) to produce high-octane gasoline and other hydrocarbon products. Conventional paraffin-olefin
(alkane-alkene) alkylation is an acid-catalyzed reaction, such as combining isobutylene and isobutane to isooctane.

$$(CH_3)_3CH + (CH_3)_2C{=}CH_2 \xrightarrow{\text{acid}} (CH_3)_2CHCH_2C(CH_3)_3$$

The key initiation step in cationic *polymerization* of alkenes is the
formation of a carbocationic intermediate, which can then interact
with excess monomer to start propagation. We studied in some detail
the initiation of cationic polymerization under superacidic, stable ion
conditions. Carbocations also play a key role, as I found not only in
the acid-catalyzed polymerization of alkenes but also in the polycon-
densation of arenes as well as in the ring opening polymerization of
cyclic ethers, sulfides, and nitrogen compounds. Superacidic oxidative
condensation of alkanes can even be achieved, including that of methane, as can the co-condensation of alkanes and alkenes.

Many superacid-catalyzed reactions were found to be carried out
advantageously not only using liquid superacids but also over solid
superacids, including Nafion-H or certain zeolites. We extensively studied the catalytic activity of Nafion-H and related solid acid catalysts
(including supported perfluorooctanesulfonic acid and its higher ho-

mologues), antimony pentafluoride and tantalum pentafluoride complexed to fluorographite, etc., but we ourselves did not study zeolites.

In addition to the study of reactions of hydrocarbons and their carbocationic intermediates, in my research on superacidic chemistry I also pursued other interests, such as the study of miscellaneous onium ions and derived varied reagents. Onium ions are the positively charged higher-valence (higher coordination) ions of nonmetallic elements. They can be obtained by protonation (alkylation, etc.) of their lower-valence parents or by ionization of appropriate precursors. Many onium ions are isolable as stable salts and are also recognized as electrophilic reaction intermediates. The onium ions we studied in superacidic media or as isolated salts included acyl cations, carboxonium, carboxonium, carbosulfonium, carbazonium, various oxonium, sulfonium, selenonium, telluronium phosphonium, halonium, siliconium, and boronium ions. I have reviewed these fields and my studies in some detail in two monographs entitled *Onium Ions* (written with colleagues), and *Halonium Ions*.

In Cleveland, I also continued my early fascination with organofluorine compounds and their chemistry. Fluorination of organic compounds requires special techniques not usually available in the average laboratory. Reactions with the most generally used and inexpensive fluorinating agent, anhydrous hydrogen fluoride, must be carried out under pressure in special equipment because of its relatively low boiling point (20°C) and corrosive nature. It is also an extremely toxic and dangerous material to work with. "Taming" of anhydrous hydrogen fluoride was thus a challenge.

We found a simple way to carry out anhydrous hydrogen fluoride reactions at atmospheric pressure in ordinary laboratory equipment (polyolefin or even glass) by using the remarkably stable complex formed between pyridine and excess hydrogen fluoride. HF (70%) and pyridine (30%) form a liquid complex, $C_5H_5NH^+(HF)_xF^-$, showing low vapor pressure at temperatures up to 60°C. This reagent (pyridinium polyhydrogen fluoride, sometimes called Olah's reagent) thus enables one to carry out a wide variety of synthetically very useful fluorination reactions safely and under very simple experimental conditions.

We also developed a number of other useful new fluorinating reagents. They included a convenient in situ form of sulfur tetrafluoride in pyridinium polyhydrogen fluoride, selenium tetrafluoride, and cyanuric fluoride. We introduced uranium hexafluoride (UF_6), depleted from the U-235 isotope, which is an abundant by-product of enrichment plants, as an effective fluorinating agent.

Studying alkylations, we developed a series of selective ionic alkylating agents. Although Meerwein's trialkyloxonium and dialkoxycarbenium salts are widely used as transfer alkylating agents, they lack selectivity and generally are incapable of C-alkylation.

In contrast, dialkylhalonium salts such as dimethylbromonium and dimethyliodonium fluoroantimonate, which we prepared from excess alkyl halides with antimony pentafluoride or fluoroantimonic acid and isolated as stable salts (the less-stable chloronium salts were obtained only in solution), are very effective alkylating agents for heteroatom compounds (Nu = R_2O, R_2S, R_3N, R_3P, etc.) and for C-alkylation (arenes, alkenes).

$$2\,RX \xrightarrow[\text{or } SbF_5]{HSbF_6} R\overset{+}{X}R + SbF_6^-$$

$$R = CH_3, C_2H_5, \text{etc.} \qquad X = I, Br, Cl$$

Because the nature of the halogen atom can be varied, these salts show useful selectivity in their alkylation reactions. We also prepared other halonium ions and studied their alkylating ability.

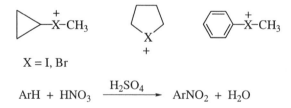

$$X = I, Br$$

$$ArH + HNO_3 \xrightarrow{H_2SO_4} ArNO_2 + H_2O$$

During my Cleveland years, I also continued and extended my studies in nitration, which I started in the early 1950s in Hungary. Conventional nitration of aromatic compounds uses mixed acid (mixture of nitric acid and sulfuric acid). The water formed in the reaction dilutes the acid, and spent acid disposal is becoming a serious environmental problem in industrial applications. Furthermore, because of its strong oxidizing ability, mixed acid is ill-suited to nitrate many sensitive compounds. We developed a series of efficient new nitrating agents and methods to overcome these difficulties. The use of readily prepared and isolated stable nitronium salts, such as $NO_2^+BF_4^-$, which I started to study in Hungary, was extended, and they became commercially available. These salts nitrate aromatics in organic solvents generally in close to quantitative yields, as well as a great variety of other organic compounds.

Some of the nitration reactions we studied with NO_2^+ salts were the following.

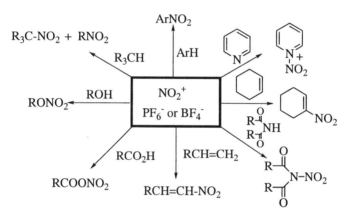

We also found N-nitropyridinium salts such as $C_5H_5N^+NO_2BF_4^-$ as convenient transfer nitrating reagents in selective, clean reactions. Transfer nitrations are equally applicable to C- as well as to O-nitrations, allowing, for example, safe, acid-free preparation of alkyl nitrates and polynitrates from alcohols (including nitroglycerine).

To solve some of the environmental problems of mixed-acid nitration, we were able to replace sulfuric acid with solid superacid catalysts. This allowed us to develop a novel, clean, azeotropic nitration of aromatics with nitric acid over solid perfluorinated sulfonic acid catalysts (Nafion-H). The water formed is continuously azeotroped off by an excess of aromatics, thus preventing dilution of acid. Because the disposal of spent acids of nitration represents a serious environmental problem, the use of solid acid catalysts is a significant improvement.

In Cleveland office, 1976. The pictures on the wall are of Meerwein, Ingold, Winstein, Brown, and Whitmore.

All in all, my Cleveland years were most rewarding. Judy was able to rejoin me in our research, and I succeeded in building up a fine research group. In a relatively short time, our efforts with my faculty colleagues resulted in creating a strong, well-balanced chemistry department in a newly emerging and dynamic university. I felt a strong

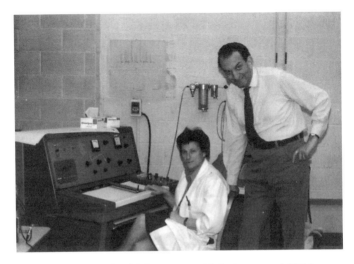

In the Cleveland laboratory with Judy around 1976

attachment to my university and my colleagues, some of whom I had recruited to come to Cleveland. By the mid-1970s, however, I began to realize that what I had set out to achieve was basically accomplished. Chemistry at CWRU was well established, and my moving on would not severely affect it. Indeed, over the years CWRU's chemistry department has continued to be strong and dynamic, giving me great satisfaction that I had a role in shaping it.

· 8 ·

Moving to Los Angeles:
Building the Loker Institute—Hydrocarbons and Hydrocarbon Research

We enjoyed our years in Cleveland, with its quiet surroundings in which to bring up our two sons. Judy was also able to rejoin me in our research. Because our home in Shaker Heights was only about a 10-minute drive from the university, it was possible to go home for lunch, and Judy miraculously managed to keep our family going with two growing boys in the house and a husband, as usual, preoccupied with his professional life. However, after 12 years, in 1977 it was time to move on. Our younger son, Ron, was finishing high school, and our older son, George, was a senior at Case Western Reserve, majoring in accounting. Ron set his heart on going to Stanford and told us for quite a while about the wonderful life in California. He felt it would be nice for the whole Olah family to resettle there.

Our natural inclination to stay put gradually weakened. Case Western Reserve University had given me a wonderful opportunity to return to academic life. I enjoyed the friendly atmosphere, my colleagues and students, and the challenge of helping to build a good chemistry department in an evolving university. However, it became increasingly clear that, despite all its positive aspects, the scope of my work in Cleveland had probably reached its limits. The excellent medical school at Case Reserve together with the neighboring Cleveland Clinic represented (and continues to do so) a very strong center for biomedical research. Similarly, the school of engineering and its programs in polymers and materials had gained well-deserved national recognition. My own interests and what I felt was the direction in which I wanted to extend my chemistry, i.e., the broad general field of hydrocarbon chemistry, however, did not seem to fit into these programs.

When I came to Cleveland as chairman of the chemistry department, Howard Schneiderman, a dynamic biologist, headed his department. Howard had foreseen the explosive development of modern biology and tried to build a significant but perhaps too ambitious biology center. He soon realized that his plans were outgrowing the possibilities offered by a moderately sized private Midwestern university. He left for the new campus of the University of California at Irvine, where as dean he built a strong biological sciences program. Subsequently he moved on to Monsanto in St. Louis, where as vice president for research he helped to transform Monsanto into a biotechnology company and built a campuslike large (maybe too large) research complex. My own ambitions were much more modest, and I was never tempted to give up my active academic career even though I was offered some interesting industrial opportunities over the years.

To move to another university was also something I did not consider seriously for a long time. There was some talk about Princeton before my friend Paul Schleyer decided to move from Princeton to Germany and some about UCLA after Saul Winstein's untimely death, but nothing developed. This had also been the case, when, unbeknownst to me (I read about it only recently in a book written about him by one of his former students, Ken Leffer), Christopher Ingold recommended me in the mid-1960s as his successor at University College of the University of London. In the early 1970s, King's College, also of the University of London, offered me its Daniell Chair in Chemistry, but I decided to stay in Cleveland. I had no intention of leaving my adopted country, which had offered me and my family a new life and home.

In the fall of 1976 I had a call from a friend, Sid Benson, who, after a decade at the Stanford Research Institute, just returned to the University of Southern California (USC) in Los Angeles. He invited me for a visit, telling me about USC's plans to build up selected programs, including chemistry. I visited USC and found it, with its close to downtown urban campus, quite different from the sprawling expanse of the cross-town campus of UCLA, which I had visited on a number of occasions during Saul Winstein's time. I knew little about USC, except that my family used to watch the New Year's Day Rose Bowl parade and football game (in which USC was a frequent participant) on TV. We had marveled at the sunny, summerlike weather of Southern California while we were freezing in snowbound Cleveland.

At this point, USC wanted to shed some of its football school image, or at least this was the goal of its dynamic and foresighted executive vice president Zohrab Kaprielian, who was then the driving force at USC. Kaprielian was an Armenian-born electrical engineer who got his education at Caltech and devoted his life to improving USC. He first built a fine electrical engineering department as its chair and followed by building up the School of Engineering as its Dean. When he became executive vice president (the equivalent of provost) under President Jack Hubbard, he launched a major effort to raise USC into the rank of major research universities. He was very persuasive in telling me about his plans but at the same time realistic about the obvious limitations. No university since Stanford's remarkable achievement in the 1950s and 1960s had been in a position to make significant improvements across the whole academic spectrum. Kaprielian thus focused on a number of areas in which he felt an impact could be made. He believed that if he could create some "centers of excellence," they would subsequently help to raise the level of the whole university. He was looking for people who he felt had promise and drive as well as a willingness to take a chance on building up something from modest beginnings.

We hit it off well. I liked his approach and his outspoken honesty about his goals. He also must have seen something in me that he liked, and we started seriously discussing my coming to USC. There were, however, difficulties. The chemistry department of USC, was (and still is) heavily centered around chemical physics (at the time, concentrating on spectroscopy). Organic chemistry for whatever reasons was much of a stepchild, although Jerry Berson spent a decade at USC before he left for Wisconsin and then Yale. Ivar Ugi was also a faculty member for three years (1969–1971) before returning to Germany. He laid the foundation of multicomponent synthesis (i.e., combinatorial chemistry) while at USC, although it did not attract much attention at the time.

The facilities for experimental work were poor, with inadequate wet laboratory space. In our discussions, I mentioned to Kaprielian my interest in significantly extending my previous work into the area of hydrocarbon chemistry. I felt that by establishing a strong program of basic research and graduate education in hydrocarbon chemistry, USC could become a leader in this important field. Because the memory of the first Arab oil embargo was still fresh, this struck a chord with Kaprielian, who felt that he could "sell" my research interest to the trustees and establish a "Hydrocarbon Research Institute" at USC that could accommodate me, as well as other chemistry faculty members whose interests could fit into its framework.

I got much advice and encouragement during my deliberations over moving to USC from Martin Kamen, a friend and true Renaissance man, as well as an outstanding scientist (co-discoverer of the carbon-14 isotope and a leading biological chemist). Martin was then attempting to build up the molecular biology program at USC. His plans never really worked out, but, characteristically, he looked at the broader picture and encouraged me. He also taught me that, this close to Hollywood, not everything is as it seems or as it is promised, and thus it is useful to have matters properly clarified and put into writing.

USC was also attempting to attract from the University of London Franz Sondheimer, a fine organic chemist whose research interest in annulenes was also in the broad scope of hydrocarbon chemistry. The two of us, together with Sid Benson, whose physicochemical research rounded out our developing plans, seemed to form an attractive group. Benson enthusiastically supported the effort and so did Jerry Segal, who as department head eased the way and solved many problems. In December 1976 both Franz Sondheimer and I agreed to come to USC. My wife and I bought a house that same month in a canyon on top of Beverly Hills, on the same street where the Sondheimers decided to live. At my son Ron's urging, we built a swimming pool, which we have used ever since for our daily morning exercise and greatly enjoy. Regrettably, Franz, who in some ways was a troubled man, changed his mind twice and eventually withdrew. Some years later he tragically ended his life, cutting short a remarkable career that much enriched chemistry.

Whenever I make up my mind about something, I never look back and move forward determinedly. Once I decided to join USC, I planned to move as soon as the coming spring, despite the fact that there were no adequate laboratory facilities available to accommodate the group of about 15 graduate students and postdoctoral fellows who decided to make the move west with me. Kaprielian put together a plan to build our own modest institute building of some 17,000 sq. ft., which was approved by the trustees after I had a chance in April 1977 to address them at their meeting in Ojai, a pleasant resort town not far from LA. It was promised to be ready for occupancy in 2 years, and the university committed the initial funds to finance it. In the meantime, some temporary space was found for us in the basement of what was called the "old science building" (which lived up to its name). It took some faith to accept this arrangement, particularly because I was determined that our move should not interrupt the research effort of my group. My wife Judy, who is more practical than I am, had serious doubts that the schedule would work out, but eventually it did. In December 1979, two and a half years after our move, we were able to occupy the new Hydrocarbon Research Institute building.

By a mutually agreeable arrangement with Case Western I was able to take most of my laboratory equipment, chemicals, and instrumentation with me to LA. I again got help from Dow Chemical, which donated an additional NMR spectrometer. We packed up our laboratories in Cleveland in late May, with my graduate students John Welch and Surya Prakash spearheading and organizing the effort. Our small convoy, including moving vans, started out across the country to California in the best spirit of westward migration. Most of my group drove to LA, but Judy and I decided to fly out with our cocker spaniel Jimmy in the baggage compartment. My sons drove out in our car, getting to know each other better during the long trip than they had in years.

The arrival of the Olah group at the USC campus with its moving vans caused quite a stir. Whatever shortcomings our temporary quarters had were overcome by our enthusiasm, and, miraculously, in 3 weeks we were back doing research. I am not sure whether everybody at USC was pleased by this "invasion" and our determination to overcome any difficulties to get started. My background may have prepared me to manage as well as possible in whatever circumstances I found myself, but it was the enthusiasm and hard work of my students that made the nearly impossible possible. I must say, however, that it was not an easy start. My group found itself isolated in many ways and left to its own resources. Organic chemists at USC were still considered outsiders. Furthermore, there is perhaps something in human nature that gives some people satisfaction to see others struggle and maybe even fail. In our case, failure would have justified the assumption by some that at USC it was not possible to succeed in organic chemistry. The Germans have a word "Schadenfreude" to express such feelings (meaning the joy of misfortune of others). Probably it would have been easier to accept us if it had taken a much longer struggle for us to settle in. In any case, we persevered and succeeded.

Early in my career, someone told me that if ever I became even mildly successful there would always be those who would envy me. I was also told, however, that this was still better than if they felt sorry for my failure or misfortune. I always remembered this advice, and it helped me to understand human nature and not to be easily offended. Universities also have their own internal politics, disagreements, and differing views. Academic infighting, as it is said, can be fierce because the stakes are so small. I always felt, however, that it is much wiser to stay out of it as much as possible and to use my energy for much more rewarding efforts in research, scholarly work, and teaching. Thus, once I realized that it would be very difficult to change the mind and views of some of my new faculty colleagues, I simply went on to pursue my own work without arguing or imposing on others. I am grateful for the help and understanding I received at USC from many. However, bringing about what I considered (and in some way still consider) necessary changes proved to be difficult. I therefore put my efforts single-mindedly toward building the Hydrocarbon Institute into a viable entity within the university with a fair degree of independence to pursue its research and educational goals. We also set our own standards.

As I mentioned, USC provided some start-up funds for the Institute. Further development and operation, however, was only possible through raising support. I never had been seriously involved in such an effort previously, but I consider myself a reasonably fast learner. During the previously mentioned trustee meeting at Ojai in the spring of 1977, Judy and I met with the leaders of the "Trojan family" (as the USC community is known), including many wonderful people. Carl Franklin, the legal vice president at the time, and his wife Caroline were particularly kind to us. The Franklins became friends, and Carl became a major supporter of our efforts.

Soon after our move to Los Angeles, Carl Franklin introduced me to Katherine and Don Loker, friends of his and benefactors of USC (as well as of other institutions including Harvard University). Katherine, a native Angelina, was a graduate of USC, where she was also a star athlete. Her father had emigrated from the Adriatic coast of Croatia to California, eventually building up a large tuna fishing and canning company (Starkist). Don was a New Englander from Boston, who after

Katherine and Donald Loker and me around 1985

Harvard moved out west and became an accomplished actor (under his professional name of Don Winslow) with some 50 movies to his credit. He later joined Starkist and became a successful businessman. Katherine and Don were married before World War II. Don's military service was in the Pacific, where he ended up on the staff of General MacArthur in Tokyo and helped to reestablish the Japanese fishing industry. During his stay in Tokyo he also learned Japanese quite well. I remember, some years later, when Reiko Choy became my secretary (a wonderful woman of Japanese origin, who for many years kept my office operating smoothly and also managed miraculously to turn my terrible hand-written papers and manuscripts, including that of this book, into proper form) that Don Loker walked in and started to talk to her in Japanese (quite a surprise coming from a New England Yankee). I still write my manuscripts in long hand, to the annoyance of my fully computer-adept wife, even refusing to use a word processor. I believe that this slow and tedious process gives me more time to review my thoughts. In any case, I may miss some wonderful technical progress, but it still works for me.

The Lokers had no background or business interest in chemistry. It was Carl Franklin who told them about USC's effort to establish a hydrocarbon research institute, and he must have also told them some-

With Carl Franklin

thing kind about me. Don's first reaction, as I understand it, was "What the hell are hydrocarbons?" Nevertheless, they eventually became interested and in 1978 made their first gift to the Institute, which in 1983 was renamed to honor them as the "Loker Hydrocarbon Research Institute." In 1991, USC, under the foresighted leadership of its new president, Steven Sample, formally adopted "Guidelines" concerning the Institute's organization and its permanence. The Lokers continued their generosity toward the Institute, which, together with that of some other friends and supporters, particularly of Harold Moulton, was invaluable to our efforts. Their friendship and sound advice, however, were equally valuable. Don Loker passed away in 1989, but Katherine continues her support and chairs the Institute's Advisory Board.

My active life and work were unexpectedly affected by two serious illnesses that I fought in 1979 and 1982. We spent the first two years in LA in temporary laboratories in the basement of an old science building, where I also had my office on the ground floor above the

With friends and benefactors, Harold Moulton and the Lokers

laboratories. Jerry Segal occupied an adjacent office. I mention this because in early 1979 I became seriously ill with a rather mysterious disease that was difficult to diagnose but that increasingly worsened and eventually became life threatening. It turned out to be a rare immune disease, which manifested itself in bad sores breaking out all over my body, which did not heal and reached frightening levels. I was seen by many dermatologists in LA (probably a mistake, because the problem was much deeper than a skin problem), but none of them could find the cause. Without realizing any connection at the time, Jerry Segal also started to show the same symptoms. By August, I was in such poor health that Judy decided as a last resort that we should go to the Mayo Clinic in Rochester, MN. This probably saved my life. Within 24 hours after our arrival the Mayo doctors diagnosed my problem as an acute case of pemphigus vulgaris and started to treat me with megadoses of Prednisone and other rather powerful medications. We planned to be at the Mayo Clinic only for a short stay for diagnosis, but I was hospitalized there for a month. Pemphigus is an immune disease, of which not much was known at the time, except that it can be triggered by certain chemicals such as penicillamine. As it turned out, in the 1940s and 1950s USC's old science building was used for pharmacological research using penicillin (and probably penicillamine). Some old air ducts of the building could still have been contaminated. Pemphigus affects only about one in a million in the United States; thus it is a true orphan disease, receiving little interest. The random probability that two individuals having adjacent offices in the same building should come down with it simultaneously is astronomical.

In any case, I eventually recovered (and so did Jerry), but my immune system must have suffered serious damage, which manifested itself three years later, when I collapsed in my office one day and was found to be bleeding internally from a form of rare stomach cancer, which necessitated major surgery but was fortunately localized. I again recovered and have had no further difficulties since. Whether weakening and knocking out my immune system to overcome the previous problems had any effect is not clear, but it could have been a factor. Despite my health problems I was able to continue my work without much interruption, and the scientific productivity of my group has not suffered. As a matter of fact, the early 1980s were some of our best years.

When you go through serious health problems and face your mortality, your attitude changes and you learn to appreciate what is really meaningful and what is only peripheral in your life. It certainly affects your general views, but it also focuses you better in your scientific work. You gain a clearer understanding of what in your work may have a more lasting value and therefore is worth pursuing, instead of scattering your efforts.

Our institute building was designed in the fall and winter of 1977, working with the noted LA architect William Pereira and his firm. Jerry Segal and Tony Lazzaro, USC vice president in charge of facilities, were most helpful in the process. The institute was given a central location in the middle of the campus adjacent to other science and engineering buildings. It replaced some World War II-vintage barracks still used for smaller classes and storage. The design provided a functional building, well suited for chemical research but at the same time also an attractive home for faculty, students, and staff. We were working with a small budget ($1.7 million), which translated to the cost of $100/sq. ft. in determining the size of the building. Obviously, much needed to be

Loker Hydrocarbon Research Institute at USC

Dedication of the new Institute building 1979—Don and Katherine Loker

economized, but the plans allowed for future expansion. The University's initial financial commitment was subsequently supplemented by a generous gift of the Lokers that allowed the building to open its doors in late 1979.

As the Institute's work and scope grew, so did the need for additional space. In 1989, Katherine Loker continued her tradition of generous giving with a major gift, which allowed the construction of a new wing (named the Katherine Bagdanovich Loker Wing). The new wing houses additional state-of-the-art laboratories, instrument facilities, and what became the George and Judith Olah Library and conference room. It contains my extensive library and periodicals collection, which we donated in 1977 when we came to USC and which we have expanded and maintained since. Additional gifts, particularly by the late Harold E. Moulton, a generous supporter and a founding member of our board, greatly advanced the Institute's development. Carl Franklin, by then emeritus vice president of USC, remained a strong supporter, and has been relentless in pursuing support for the Institute. We are extremely grateful to them and to others for their generosity and support.

The extended Institute now has some 43,000 sq. ft. of space in a most attractive setting and provides not only first-class research facilities but in a sense a home away from home for all in the Loker Institute family, including what the younger generation named "le bistro," a pleasant setting for lunch and informal discussions.

The new addition of the building was completed by the end of 1994 and dedicated in February 1995. Because I coincidentally won the Nobel Prize just two months before (more about this in Chapter 11), some believed that there was some relationship between the two events. This certainly was not the case. Katherine Loker and our other friends had made their wonderful gifts well before, and it was just a fortunate coincidence that we had such "good timing" to celebrate the opening of our enlarged institute.

Physical facilities help, but do not per se make a research institute. It is the people who work there and their contributions and devoted hard work that is most important. We are nearing a quarter of a century since the Hydrocarbon Research Institute was started at USC. At the beginning in 1977, Sid Benson and I shared the scientific directorship of the Institute and Jerry Segal carried out the administrative responsibilities as executive director. When we moved into our own building in 1979, Bill Stephenson, a physical-organic chemist and a former colleague of mine in Cleveland who subsequently joined us at

Katherine Loker wing of the Institute

The Olah Library

Dedication of the Olah Library with Katherine Loker in 1996

With Judy at the Library Dedication, 1996

USC, took over as executive director. Bill, a fine scientist and teacher, was also an outstanding colleague and good administrator. In 1983, however, he decided to pursue other interests and left academia.

To fill the void, we attracted another USC colleague John Aklonis, a polymer physicochemist, as executive director. With John joining the Institute, we made a commitment to include polymer chemistry and, subsequently, materials research as an integral effort. John became a close friend and is one of the nicest individuals I have ever met. He contributed greatly to our efforts and also helped me personally to continue to stay active in my research by taking most of the burden of administrative duties off me. His outgoing personality and humanity made him loved and respected by everyone and allowed the Loker Institute to continue and expand its work. John should be given the lion's share of credit for making the expansion of our Institute a reality, helping to raise funds, planning the addition to our building, and keeping the Institute afloat in the sometimes turbulent waters of the internal politics of the university. John and his wife Jo were always avid sailors. Their life was very much centered around their sailing, and they even lived on their boat. In 1990 John, approaching the age of 50, decided to give up his university career, take early retirement, and set sail to spend the next decade (or two) on the high seas. Ten years later, after having sailed halfway around the world, they still are at it in the Far East, but I guess some day they will come back to the United States. We miss them very much, but we keep in touch and they fly back to visit LA occasionally. We wish them continued enjoyment, satisfaction, and safe sailing. From the Institute's and my own point of view, however, John's departure was a serious blow, and he was not easily replaced. Somewhat reluctantly, I took on the overall duties as director. Bill Weber and Robert Aniszfeld, who was a graduate student of mine, became our excellent associate directors, and together with our small but highly dedicated and efficient staff they take care of most administrative duties.

In shaping the scientific work of the Loker Institute as well as pursuing my own research interests, I got invaluable help from the colleagues who joined in the effort, and particularly from G. K. Surya Prakash, my former graduate student who stayed on. Over the years he became a highly respected professor, my partner in our joint research, and a close friend. It is not often that outstanding scientific ability and wonderful human characteristics are combined in a person. Prakash is such a unique person. I am lucky to have been associated with him for a quarter of century. I am also proud to have seen him develop into an outstanding scholar and teacher in his own right. He moved up in his academic career to a professorship, and when Judy and I donated part of my Nobel money to establish, together with the generous help of Carl Franklin and other friends, a professorship at USC, to my delight, Prakash was named the first "George and Judy Olah Professor." He is now also scientific codirector of the Loker Institute and doing a great job. Other faculty colleagues in our Institute include in the organic/polymer area Bill Weber, Thieo Hogen-Esch, Nicos Petasis, and more recently Roy Periana and Aaron Harper. Golam

With Surya Prakash

Rasul, a research faculty member, contributes significantly to our computational chemistry efforts. Bob Williams continues his fascinating probing of boron compounds. Karl Christe, an outstanding inorganic chemist and an adjunct professor, heads an extremely productive laboratory in the chemistry of fluorine compounds and high-energy compounds. Bob Bau provides all of us with great help in x-ray crystallography. After Sid Benson's retirement, Larry Dalton (also a scientific co-director) Jim Haw and Marc Thompson have spearheaded efforts in the more physically oriented materials-catalysis area.

Since its inception, the institute, in addition to its regular faculty members, has always had a number of outstanding scientists associated with it as fellows or adjunct faculty. These include, among others, Ned Arnett, Joe Casanova, Paul Schleyer, Jean Sommer, Gabor Somorjai, Peter Stang, Michael Szwarc (who just recently passed away), Ken Wade, and Bob Williams. They greatly contribute to our scientific work by their visits and participation in our symposia and frequently through close research cooperation. Because they are also close friends,

With Paul Schleyer

our association adds much to the enjoyment and pleasure of our efforts. I only hope that they also feel the same in some way.

The objectives of the Loker Hydrocarbon Research Institute have stayed the same since its inception in 1977. They are:

- ♦ To pursue the long-range development of hydrocarbon chemistry
- ♦ To develop new fuels and materials and to provide environmentally sustainable solutions to energy generation problems
- ♦ To train researchers in the field of hydrocarbon chemistry
- ♦ To further the interchange of information through publications and symposia on developments in hydrocarbon chemistry
- ♦ To act as an international center of hydrocarbon chemistry and to facilitate exchange of information and ideas through visits of scientists, colloquia, and research symposia

With Peter Stang

◆ To foster a close relationship with the chemical, petroleum, gas, and
   energy industries and governmental agencies for the exchange of information and knowledge and to ensure that research results and discoveries of significance will be effectively exploited

Hydrocarbons, as their name indicates, are compounds of carbon
and hydrogen. They represent one of the most significant classes of
organic compounds. The scope of hydrocarbons is broad. They include
saturated hydrocarbons (alkanes, cycloalkanes) and their derivatives as
well as unsaturated alkenes and dienes, acetylenes, and aromatics. In
methane (CH_4), the simplest saturated alkane, a single carbon atom is
bonded to four hydrogen atoms by sharing electron pairs, i.e., by covalent bonds. In the higher homologues of methane (of the general
formula C_nH_{2n+2}), all atoms are bound to each other by such single
[sigma (σ), two electron-two center] bonds, with carbon atoms also
displaying their ability bind to each other to form C-C bonds. Carbon
atoms can be aligned in open chains (acyclic hydrocarbons). Whereas
in CH_4 the H:C ratio is 4, in C_2H_6 (ethane) it is decreased to 3, in
C_3H_8 (propane) to 2.67, and so on. Alkanes can be straight-chain (each

carbon attached to 2 other carbon atoms) or branched (in which at least 1 carbon is attached to either 3 or 4 other carbon atoms).

Carbon can also form multiple bonds with other carbon atoms. This results in unsaturated hydrocarbons such as olefins (alkenes), containing a carbon-carbon double bond, or acetylenes (alkynes), containing a carbon-carbon triple bond. Dienes and polyenes contain two or more unsaturated bonds.

Carbon atoms can also form cyclic compounds. Aromatic hydrocarbons (arenes), of which benzene is the parent, consist of a cyclic arrangement of formally unsaturated carbons, which, however, give a stabilized (in contrast to their hypothetical cyclopolyenes), delocalized system.

The H:C ratio in hydrocarbons is indicative of the hydrogen deficiency of the system. As mentioned, the highest theoretical H:C ratio possible for hydrocarbon is 4 (in CH_4), although in electron-deficient carbocationic compounds such as CH_5^+ and even CH_6^{2+}, the ratio is further increased (to 5 and 6, respectively, see Chapter 10). On the other end of the scale in extreme cases, such as the dihydro- or methylene derivatives of recently discovered C_{60} and C_{70} fullerenes, the H:C ratio can be as low as 0.03.

Hydrocarbons are abundant in nature. All *fossil fuels* (coal, oil, gas) are basically hydrocarbons, deviating, however, significantly in their H:C ratio.

H:C ratio of natural hydrocarbon sources

Methane	4.0
Natural Gas	3.8
Petroleum crude	1.8
Tar sands bitumen	1.5
Shale oil (raw)	1.5
Bituminous coal	0.8

Natural gas, depending on its source, contains—besides methane as the main hydrocarbon compound (present usually at >80–90%)—some of the higher homologous alkanes (ethane, propane, butane). In "wet" gases the amount of C_2-C_5 alkanes is higher (gas liquids).

Petroleum or *crude oil* is a complex mixture of many hydrocarbons. It consists of saturated, predominantly straight-chain alkanes, small amounts of slightly branched alkanes, cycloalkanes, and aromatics. Petroleum is generally believed to be derived from organic matter deposited in the sediments and sedimentary rocks on the floor of marine basins. The identification of biological markers such as petroporphyrins provides convincing evidence for the biological origin of oil (however, there may also be hydrocarbons of abiogenic origin). The effects of time, temperature, and pressure in the geological transformation of the organics to petroleum are not yet clear. However, considering the low level of oxidized hydrocarbons and the presence of porphyrins, it can be surmised that the organics were acted upon by anaerobic microorganisms and that temperatures were moderate, <200°C. By comparing the elemental composition of typical crude oils with typical bituminous coals, it becomes clear why crude oil is a much more suitable fuel source in terms of its higher H:C atomic ratio, generally lower sulfur and nitrogen contents, very low ash contents (probably mostly attributable to suspended mineral matter and vanadium and nickel associated with porphyrins), and essentially no water content.

It is interesting to note that recent evidence shows that even extra-terrestrially formed hydrocarbons can reach the Earth. The Earth continues to receive some 40,000 tons of interplanetary dust every year. Mass-spectrometric analysis has revealed the presence of hydrocarbons attached to these dust particles, including polycyclic aromatics such as phenanthrene, chrysene, pyrene, benzopyrene, and pentacene of extra-terrestrial origin indicated by anomalous isotopic ratios.

Petroleum—a natural mineral oil—was referred to as early as the Old Testament. The word *petroleum* means "rock oil" [from the Greek *petros* (rock) and *elaion* (oil)]. It has been found for centuries seeping out of the ground, for example, in the Los Angeles basin in what are now called the La Brea tar pits. Vast deposits were found in Europe, Asia, the Americas, and Africa.

In the United States, the first commercial petroleum deposit was discovered in 1859 near Titusville in western Pennsylvania when Edwin Drake and Bill Smith struck oil in their first shallow (~20 m) well. The well yielded some 400 gallons of oil per day (about 10 barrels).

The area had already been known to contain petroleum, which residents skimmed from the surface of a local creek, called "Oil Creek." The first oil-producing well rapidly opened up a whole new industry. The discovery was not unexpected, but it provided evidence for oil deposits in the ground that could be reached by drilling. Oil was used for many purposes at that time, such as burning in lamps and even medical remedies. The newly discovered Pennsylvania petroleum was soon also marketed to degrease wool, prepare paints, fuel steam engines, power light railroad cars, and for many other uses. It was recognized that the oil was highly impure and had to be refined to separate different fractions for varied uses. The first petroleum refinery, a small stilling operation, was established in Titusville in 1860 and subsequently John D. Rockefeller started his Standard Oil empire with a refinery in his hometown of Cleveland. Petroleum refining was much cheaper than producing coal oil (kerosene), and soon petroleum became the predominant source for kerosene as an illuminant. With the popularity of automobiles in the 1910s, gasoline became the major petroleum product. Large petroleum deposits were found in California, Texas, Oklahoma, and, more recently, Alaska. Areas of the Middle East, Asia, Russia, Africa, South America, and, more recently, the North Sea became major world oil production centers.

The daily consumption of crude oil in the United States is about 18–20 million barrels. The world consumption is about 75–80 million barrels (some 10–12 million tons) per day; the United States uses about 18–20% of the world's total but has less than 5% of the world population. Most of this oil is used for the generation of electricity, for heating, and as transportation fuel. About 4% of the petroleum and natural gas is used as feedstocks for the manufacturing of chemicals, pharmaceuticals, plastics, elastomers, paints, and a host of other products. Petrochemicals from hydrocarbons provide many of the necessities of modern life, to which we have become so accustomed that we do not even notice our increasing dependence, even though the consumption of petrochemicals is still growing at an annual rate of 10%. Advances in the petroleum-hydrocarbon industry, more than anything else, may be credited for the high standard of living we enjoyed in the late twentieth century.

Whereas light crudes are preferred in present-day refining operations, increasingly, *heavy petroleum sources* also must be processed to satisfy ever-increasing needs. These range from commercially usable heavy oil (California, Venezuela, etc.) to the huge petroleum reserves locked up in *shale* or *tar sand* formations. These more unconventional hydrocarbon accumulations exceed the quantity of lighter oil present in all the rest of the oil deposits in the world together. One of the largest accumulations is located in Alberta, Canada, in the form of large tar sand and carbonate rock deposits containing some 2.5–6 trillion barrels of extremely heavy oil called *bitumen*. There are large heavy oil accumulations in Venezuela and Siberia, among other areas. Another vast, commercially significant reservoir of oil is the oil shale deposits located in Wyoming, Utah, and Colorado. The practical use of these potentially vast reserves will depend on finding economical ways to extract the oil (by thermal retorting or other processes) for further processing. Alberta tar-sand oil is already processed in commercially viable large-scale operations.

The quality of petroleum varies, and, according to specific gravity and viscosity, we talk about light, medium, heavy, and extra heavy crude oils. Light oils of low specific gravity and viscosity are more valuable than heavy oils with higher specific gravity and viscosity. In general, light oils are richer in saturated hydrocarbons, especially straight-chain alkanes, than are heavy oils; they contain <75% straight-chain alkanes and <95% total hydrocarbons. Extra heavy oils, the bitumens, have a high viscosity, and thus may be semisolids with a high heteroatom content (nitrogen, oxygen, and sulfur) and a corresponding reduced hydrocarbon content, of the order of 30–40%.

Typical percentage compositions of light and heavy oils are given below.

Fraction	Light Oil	Heavy Oil
Saturates	78	17–21
Aromatics	18	36–38
Resins	4	26–28
Asphaltene	Trace–2	17

Heavy oils and especially bitumens contain high concentrations of resins (30–40%) and asphaltenes (<20%). Most heavy oils and bitumens are thought to be derivatives of lighter, conventional crude oils that have lost part or all of their straight-chain alkane contents along with some of their low-molecular-weight cyclic hydrocarbons through processes taking place in the oil reservoirs. Heavy oils are also abundant in heteroatom (N, O, S)-containing molecules, organometallics, and colloidally dispersed clays and clay organics. The prominent metals associated with petroleum are nickel, vanadium [mainly in the form of vanadyl ions (VO^{2+})], and iron. Some of these metals are (in part) bound to porphyrins to form metalloporphyrins.

Processing heavy oils and bitumens represents a challenge for the current refinery processes, because heavy oils and bitumens poison the metal catalysts used in the refineries. In our research at the Loker Institute, we found the use of superacid catalysts, which are less sensitive to heavy oils, an attractive solution to their processing, particularly hydrocracking.

Coals (the plural is deliberately used because coal has no defined, uniform nature or structure) are fossil sources with low hydrogen content. The "structure" of coals means only the structural models depicting major bonding types and components relating changes with coal rank. Coal is classified, or ranked, as lignite, subbituminous, bituminous, and anthracite. This is also the order of increased aromaticity and decreased volatile matter. The H:C ratio of bituminous coal is about 0.8, whereas anthracite has H:C ratios as low as 0.2.

From a chemical, as contrasted to a geologic, viewpoint the coal formation (coalification) process can be grossly viewed as a continuum of chemical changes, some microbiological and some thermal in origin, involving a progression in which woody or cellulosic plant materials (the products of nature's photosynthetic recycling of CO_2) in peat swamps are converted over many millions of years and under increasingly severe geologic conditions to coals. Coalification is grossly a deoxygenation-aromatization process. As the "rank" or age of the coal increases, the organic oxygen content decreases and the aromaticity (defined as the ratio of aromatic carbon to total carbon) increases. Lignites are young or "brown" coals that contain more organic oxygen functional groups than do subbituminous coals, which in turn have a higher carbon content but fewer oxygen functionalities.

The organic chemical structural types believed to be characteristic of coals include complex polycyclic aromatic ring systems with connecting bridges and varied oxygen-, sulfur-, and nitrogen-containing functionalities.

The main approaches used to convert coals to liquid hydrocarbons (coal liquefaction) center around breaking down the large, complex "structures," generally by hydrogenative cleavage reactions, and increasing the solubility of the organic portion. Coal liquefaction can be achieved by direct catalytic hydrogenation (pioneered by Bergius, Nobel Prize in chemistry 1931). Combinations of alkylation, hydrogenation, and depolymerization reactions followed by extraction of the reacted coals are the major routes taken. This can provide liquid fuels, providing gasoline and heating oil.

Different types of other coal liquefaction processes have been also developed to convert coals to liquid hydrocarbon fuels. These include high-temperature solvent extraction processes in which no catalyst is added. The solvent is usually a hydroaromatic hydrogen donor, whereas molecular hydrogen is added as a secondary source of hydrogen. Similar but catalytic liquefaction processes use zinc chloride and other catalysts, usually under forceful conditions (375–425°C, 100–200 atm). In our own research, superacidic $HF-BF_3$-induced hydroliquefaction of coals, which involves depolymerization-ionic hydrogenation, was found to be highly effective at relatively modest temperatures (150–170°C).

The ultimate "depolymerization" of coal occurs in Fischer-Tropsch chemistry, in which the coal is reacted with oxygen and steam at about 1100°C to break up and gasify it into carbon monoxide, hydrogen, and carbon dioxide. A water-gas shift reaction is then carried out to adjust the hydrogen:carbon monoxide ratio of syn-gas, after which the carbon monoxide is catalytically hydrogenated to form methanol or to build up liquid hydrocarbons. Similarly, natural gas can also be used to produce syn-gas. Because the Fischer-Tropsch chemistry is, however, highly energy demanding and uses limited and nonrenewable fossil fuel sources, other approaches are needed for the future. A significant part of our recent research effort in the Loker Institute is centered on the chemical reductive recycling of atmospheric carbon dioxide into methyl alcohol and derived hydrocarbons (see Chapter 13).

Crude oil (petroleum), a dark viscous liquid, is a mixture of virtually hundreds of different hydrocarbons. Distillation of the crude oil yields several fractions, which are then used for different purposes.

Fractions of typical distillation of crude petroleum

Boiling Point Range (°C)		
<30	C_1-C_4	Natural gas, methane, ethane, propane, butane, liquefied petroleum gas
30–200	C_4-C_{12}	Petroleum ether (C_5-C_6), ligroin (C_7), straight-run gasoline
200–300	C_{12}-C_{15}	Kerosene, heating oil
300–400	C_{15}-C_{25}	Gas oil, diesel fuel, lubricating oil, waxes
>400	>C_{25}	Residual oil, asphalt, tar

The relative amounts of usable fractions that can be obtained from a crude oil do not coincide with commercial needs. Also, the qualities of the fractions obtained directly by distillation of the crude oil also seldom meet the required specifications for various applications; for example, the octane rating of the naphtha fractions must be substantially upgraded to meet the requirements of internal-combustion engines in today's automobiles. These same naphtha liquids must also be treated to reduce sulfur and nitrogen components to acceptable levels (desulfurization and denitrogenation) to minimize automotive emissions and pollution of the environment. Therefore, each fraction must be upgraded in the petroleum refinery to meet the requirements for its end-use application. The various fractions of the refining operations are further converted or upgraded to needed products, such as high-octane alkylates, oxygenates, and polymers. Major hydrocarbon refining and conversion processes include cracking, dehydrogenation (reforming), alkylation, isomerization, addition, substitution, oxidation-oxygenation, reduction-hydrogenation, oligomerization, polymerization, and metathesis.

The hydrocarbon research program of the Loker Institute was able in many ways to build on and utilize results of our fundamental work on superacid-catalyzed reactions and their mechanistic aspects (including carbocationic intermediates) to develop practical processes. The study of such processes became one of the major areas of our research. For example, we developed an environmentally friendly and practical alkylation process for the manufacture of high-octane gasoline using additives that modify the high volatility of toxic hydrogen fluoride catalyst used industrially and allow its safe use. As mentioned above, we also found new ways of hydrocracking coal, shale oil, tar sands, and other heavy petroleum sources and residues using superacidic catalysts. Much work was also done to find new environmentally adaptable and efficient oxygenates for cleaner-burning high-octane gasoline. We also developed improved diesel fuels, making them cleaner burning with high cetane ratings, without the use of toxic additives. Although our ongoing research is in part aimed at more efficient utilization of our still-existing fossil fuel resources and development of environmentally benign new chemistry, a major emphasis is to develop new ways to produce hydrocarbons. This involves new ways of converting methane (natural gas) into higher hydrocarbons (Olah, Prakash, and Periana) and the reductive recycling of excess carbon dioxide (which is at the same time a major greenhouse gas responsible for much of global warming) to useful fuels and products (as discussed in Chapter 13).

Although the major emphasis of the work in the Loker Institute was and is directed in the broadest sense toward the study of the fundamental chemistry of hydrocarbons, substantial and increasing emphasis is also directed to the aspects of hydrocarbon transformations as well as of derived polymeric and varied synthetic materials.

As a part of a broader hydrocarbon chemistry focus, the study and development of selective *synthetic reagents and methods* are also being pursued (Petasis, Prakash). These include organometallic systems, particularly boron- and titanium-based reagents, electrochemical synthesis, and selective alkylating, fluorinating, nitrating, oxygenating, carbonylating, and other reagent systems. Emphasis is also given to asymmetric reactions and new catalytic transformations. The synthetic research effort is supplemented by mechanistic studies.

My faculty colleagues of the Institute also bring great expertise in the areas of anionic, cationic, and radical polymerization to the transformation of low-molecular-weight hydrocarbons into macromolecules, oligomers, and polymers (Hogen-Esch and Weber). Research is also actively pursued in the areas of organosilicon chemistry (Weber) and the stereochemical and topological control of supramolecular structures. The *polymer and materials chemistry* effort at the Institute (Dalton, Harper and Thompson) primarily focuses on the development of polymeric materials with novel electronic structures and new molecular architectures. Such new materials are designed for the purpose of achieving new physical and chemical properties relevant to applications ranging from photochemical energy conversion to high-speed information processing and biomedical applications. Two directions of research have proven particularly fruitful, the development of electroactive materials containing molecular segments with extended-electron conjugation and the development of nanoscale (10^{-9} meter) materials.

Electroactive polymers exhibit interesting new electrical, optical, and magnetic properties. Recent examples include metallic-like conductivity and photoconductivity, superconductivity, a wide range of new magnetic phenomena, and both linear and nonlinear optical phenomena. New light-emitting diodes, solid-state organic lasers, and electro-optic devices are but a few of the exciting optical applications that are being developed using electroactive polymeric materials. Polymeric electro-optic modulators have permitted the realization of information processing bandwidths (data handling rates) of greater than 100 GHz. This has been accomplished with electrical control voltages on the order of 1 volt and in sophisticated integrated devices consisting of organic modulators, VLSI semiconductor electronics, and silica fiber optic transmission lines. Applications of polymeric electro-optic modulators extend from the cable television industry to the video display industry, biomedical sensing, and radar technology. Nanotechnology is directed toward the construction of structures and devices whose components are measured in. *Nanochemistry* is capable of producing such components, permitting a wide range of new phenomena such as photonic bandgap phenomena to be demonstrated and exploited. Light-harvesting dendrimeric materials have been developed that permit energy to be efficiently collected over the electromagnetic radiation spectrum from the ultraviolet to the infrared and converted to a single emission wavelength. This, in turn, can be used to amplify fiber optic communication signals or to carry out chemistry analogous to the photochemistry of green plants. Block copolymers are being prepared and used to achieve a variety of new electrical and optical properties. However, our research at the Loker Hydrocarbon Institute goes beyond simply applying nanostructure concepts to specific applications; rather, an effort is carried out to develop systematic new approaches to the synthesis of a wide range of nanoarchitecture using novel synthetic methods.

An integral and equally important part of the Loker Institute's efforts is its educational work. Our research is intimately coupled with the education and development of the next generation of researchers, who will play an essential role in years to come in helping to solve the very challenging problems our society faces in production and use of hydrocarbons and their products in new, safe, and environmentally adaptable ways. You cannot separate teaching from doing de facto research. To facilitate this effort, our faculty, in addition of teaching varied courses and guiding graduate students, also welcomes undergraduates for participation in research projects. Besides our own undergraduates we have hosted over the years a series of German Adenauer Fellows doing research with us and preparing their undergraduate research theses (diploma work) in the Institute. They, including Thornston Bach, all did an outstanding job.

We also established and carry on an active seminar program and international symposia (once or twice a year) on topics of general interest in our broad field. To date, our Institute has sponsored more than 30 of these highly successful symposia, covering a wide variety of topics. They bring together leading researchers from around the world and also allow our faculty and students close personal interaction with them. The Stauffer Charitable Trust endowed these symposia some years ago, and we named them for a friend of the Institute, the late Richard Kimbrough, a leading LA attorney and long-time chair of the Stauffer Trust.

· 9 ·

"Every Scientist Needs Good Enemies":
The Nonclassical Ion Controversy and Its Significance

Controversies are an integral part of the progress of science. Many scientific controversies have been well documented. They usually are centered around challenges to established concepts or theories when new facts become available. It took observations using the telescope (in its original primitive but quite effective form) to allow Kepler and Galileo to question the central position of the Earth among the celestial bodies. The theory of the void of space being filled with "ether" succumbed to Michelson and Morley's experiment. The phlogiston theory of chemistry gave away under recognition of the atomic concept of elements. Controversies are eventually resolved because in the physical sciences "proof" (or according to Popper "falsification") can lead to answers concerning specific questions when new observations or experimental data are obtained, forcing inevitable conclusions. Of course, such fundamental questions as those of our being, the origin of the universe (or universes), of intelligent life, and free will, cannot be similarly approached. Fortunately for chemists, we do not deal with questions reaching outside the limits of knowledge of our field, and thus our controversies are eventually resolved.

One of the major contemporary chemical controversies in which I was inadvertently involved developed in the 1950s, surprisingly over the structure of a deceptively simple seven carbon-containing bicyclic carbocation, the 2-norbornyl (bicyclo[2.2.1]heptyl) cation. The involvement of the ion in the long-recognized rearrangement of norbornyl systems prevalent in natural terpenes was first suggested in 1922 by Meerwein. The remarkable facility of skeletal rearrangements in norbornyl systems attracted the early interest of chemists. Wagner realized first in 1899 the general nature of these rearrangements and related them to that which takes place during the dehydration of pinacol to tetramethylethylene. Sommer later found some tricyclanes in the products of the Wagner rearrangements of terpenes. In 1918 Ruzicka suggested a tricyclane-type mechanism without realizing the ionic nature of the process. Meerwein reconsidered the mechanism in 1922 and made the farsighted suggestion that the reaction proceeds through an ionic intermediate, i.e., the norbornyl cation. Hence this type of transformation is now known as the Wagner-Meerwein rearrangement (see p. 74).

The structure of the norbornyl cation became controversial in the "nonclassical ion" controversy following Wilson's original suggestion in 1939 of a mesomeric, σ-delocalized, carbocationic intermediate in the camphene hydrochloride-isobornyl chloride rearrangement. From 1949 to 1952, Winstein and Trifan reported a solvolytic study of the exo- and endo-2-norbornyl brosylates (p-bromobenzenesulfonates) and postulated a σ-delocalized, symmetrically bridged norbornyl ion intermediate. The endo reactant was found to solvolyze in acetic acid, aqueous acetone, and aqueous dioxane to give substitution products of exclusively exo configuration, whereas the exo-brosylate giving exclusively exo-product was markedly more reactive in acetolysis than the endo, by a factor of 350.

Winstein, one of the most brilliant chemists of his time, concluded that "it is attractive to account for these results by way of the bridged (non-classical) formulation for *the norbornyl cation involving accelerated rate of formation from the exo precursor [by anchimeric assistance]*." His formulation of the norbornyl cation as a σ-bridged species stimulated other workers in the solvolysis field to interpret results in a variety of systems in similar terms of σ-delocalized, bridged carbonium ions.

H. C. Brown (the pioneer of hydroboration chemistry, Nobel Prize, 1979), in contrast, concluded that in solvolysis both 2-exo and 2-endo norbornyl esters (brosylates, etc.) undergo anchimerically unassisted ionization and that the singular rate and product characteristics of the system are attributable to steric effects, in particular, hindrance to ionization of the *endo* isomers. Explaining the results of the extensive solvolytic studies, he suggested that high *exo/endo* rate and product ratios do not necessitate σ-participation as an explanation. In other words, *exo* is not fast; *endo* is slow. His suggestion for the structure of the norbornyl cation was that of a rapidly equilibrating pair of regular trivalent ions (classical ions) which he compared to a windshield wiper. However, at the same time none of his studies ever showed that σ-participation is not involved.

In 1962, H. C. Brown lodged his dissent against the σ-bridged 2-norbornyl cation and, for that matter, other nonclassical carbocations. He has maintained his position virtually unchanged over the years and has continued to present his views forcefully. In arguing against carbon σ-bridging, he took the position, despite his pioneering work in structurally closely related boranes, that if carbon were to participate in bridging, novel bonding characteristics must be attributed to it.

In 1965 he stated, "On the other hand, the norbornyl cation does not possess sufficient electrons to provide a pair for all of the bonds required by the proposed bridged structures. *One must propose a new bonding concept, not yet established in carbon structures*" (emphasis added).

In 1967 he again wrote, "The second subclass consists of ions such as the bicyclobutonium and the norbornyl cation in its σ-bridging form, which do not possess sufficient electrons to provide a pair for all of the bonds required by the proposed structures. *A new bonding concept not yet established in carbon structures is required*" (emphasis added).

The Brown-Winstein nonclassical ion controversy can be summed up as differing explanations of the same experimental facts (which were obtained repeatedly and have not been questioned) of the observed significantly higher rate of the hydrolysis of the 2-*exo* over the 2-*endo*-norbornyl esters. As suggested by Winstein, the reason for this is participation of the C_1-C_6 single bond leading to delocalization in the bridged "nonclassical" ion. In contrast, Brown maintained that the cause was only steric hindrance to the sterically hindered *endo* side involving rapidly equilibrating "classical" trivalent ions.

Nonclassical ions, a term first used by John Roberts (an outstanding Caltech chemist and pioneer in the field), were defined by Paul Bartlett of Harvard as containing too few electrons to allow a pair for each "bond"; i.e., they must contain delocalized σ-electrons. This is where the question stood in the early 1960s. The structure of the intermediate 2-norbornyl ion could only be suggested indirectly from rate (kinetic) data and observation of stereochemistry; no direct observation or structural study was possible at the time.

My own involvement with the norbornyl ion controversy goes back to 1960–1962, when I succeeded in developing a general method of preparing and studying persistent (long-lived) alkyl cations (Chapter 6). Not unexpectedly, my interest extended to the study of various carbocations, including the controversial 2-norbornyl cation. Whereas previous investigators were able to study carbocations only indirectly (by kinetic and stereochemical studies), my newly discovered methods allowed their preparation and direct study as persistent (long-lived) species.

The 1962 Brookhaven Mechanism Conference, where I first reported on long-lived carbocations in public, is still clear in my mind. The scheduled "main event" of the meeting was the continuing debate between Saul Winstein and Herbert Brown on the classical or non-classical nature of carbocations (or carbonium ions as they were still called at the time). It must have come as a surprise to them and to the audience that a young chemist from an industrial laboratory was invited to give a major conference lecture to report having obtained and studied stable, long-lived carbonium ions (i.e., carbocations) by the new method of using a highly acidic (superacidic) system. I remember being called aside separately during the conference by Winstein and Brown, both towering and dominating personalities of the time who cautioned me that a young man should be exceedingly careful in making such claims. Each pointed out that most probably I was wrong and could not have obtained long-lived carbonium ions. Just in case, however, my method turned out to be real, I was advised to obtain further evidence for the "nonclassical" or "classical" nature (depending on who was giving the advice) of the much-disputed 2-norbornyl cation.

Because my method indeed allowed me to prepare carbocations as long-lived species, clearly the opportunity was there to experimentally decide the question through direct observation of the ion. At the time of the Brookhaven conference I had already obtained the proton NMR spectrum of 2-norbornyl fluoride in SbF$_5$, but only at room temperature, which displayed a single broad peak indicating complete equilibration through hydride shifts and Wagner-Meerwein rearrangement (well known in solvolysis reactions and related transformations of 2-norbornyl systems). However, my curiosity was aroused, and when I moved to Dow's Eastern Research Laboratory in 1964, the work was further pursued in cooperation with Paul Schleyer from Princeton (who became a lifetime friend) and Martin Saunders from Yale. Using SO$_2$ as solvent, we were able to lower the temperature of the solution to $-78°C$, and we also prepared the ion by ionization of β-cyclopentenylethyl fluoride or by protonation of nortricylene in FSO$_3$H;SbF$_5$/SO$_2$ClF. The three separate routes (representing σ-, π- or bent σ-participation) gave the identical ion.

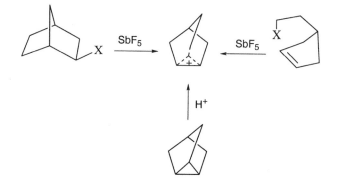

I still did not have suitable low-temperature instrumentation of my own to carry out the low-temperature NMR studies, but Martin Saunders at Yale did. Thus our samples now traveled the Massachusetts Turnpike from Boston to New Haven, where with Marty we were able to study solutions of the norbornyl cation at increasingly lower temperatures using his home-built variable-temperature NMR instrumentation housed in the basement of the old Yale chemistry building. We were able to obtain NMR spectra of the ion at −78°C, where the 3,2-hydride shift was frozen out. However, it took until 1969, after my move to Cleveland, to develop efficient low-temperature techniques using solvents such as SO_2ClF and SO_2F_2. We were eventually able to obtain high-resolution ^1H and ^{13}C NMR spectra (using ^{13}C-enriched precursor) of the 2-norbornyl cation down to −159°C. Both the 1,2,6 hydride shifts and the Wagner-Meerwein rearrangement were frozen out at such a low temperature, allowing us to observe the static, bridged ion, which I first reported at the Salt Lake City Organic Chemistry Symposium in 1969.

The differentiation of bridged nonclassical from rapidly equilibrating classical carbocations based on NMR spectroscopy was difficult because NMR is a relatively slow physical method. We addressed this question in our work using estimated NMR shifts of the two structurally differing ions in comparison with model systems. Later, this task

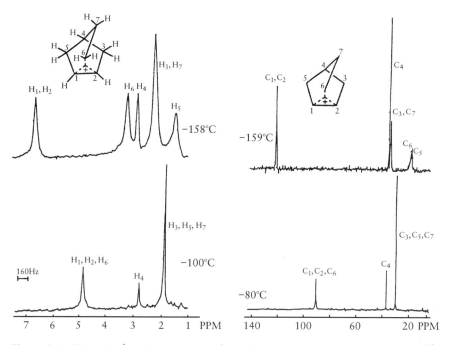

Figure 9.1. 395 MHz ^1H NMR spectra of 2-norbornyl 50 MHz proton decoupled ^{13}C NMR spectra cation in $SbF_5/SO_2ClF/SO_2F_2$ solution of 2-norbornyl cation (^{13}C enriched) in $SbF_5/SO_2ClF/SO_2F_2$ solution.

became greatly simplified and more precise by highly efficient theoretical methods such as IGLO and GIAO, allowing the calculation of NMR shifts of differing ions and comparison with experimental data. It is rewarding to see, however, that our results and conclusions stood up well against all the more recent advanced studies.

As mentioned, we also carried out IR studies (a fast vibrational spectroscopy) early in our work on carbocations. In our studies of the norbornyl cation we obtained Raman spectra as well, although at the time it was not possible to theoretically calculate the spectra. Comparison with model compounds (the 2-norbornyl system and nortricyclane, respectively) indicated the symmetrical, bridged nature of the ion. In recent years, Sunko and Schleyer were able, using the since-developed Fourier transform-infrared (FT-IR) method, to obtain the spectrum of the norbornyl cation and to compare it with the theoretically calculated one. Again, it was rewarding that their data were in excellent accord with our earlier work.

Kai Siegbahn's (Nobel Prize in physics, 1981) core electron spectroscopy (ESCA) was another fast physical method that we applied to further resolve the question of bridged versus rapidly equilibrating ions. We were able to study carbocations in the late 1960s by this method, adapting it to superacidic matrixes. George Mateescu and Louise Riemenschneider in my Cleveland laboratory set up the necessary methodology for obtaining the ESCA spectra of a number of carbocations, including the *tert*-butyl and the 2-norbornyl cation in SbF_5-based superacidic matrixes. These studies again convincingly showed the nonclassical nature of the 2-norbornyl cation. No trivalent carbenium ion center characteristic of a "classical" ion, such as is the case for the *tert*-butyl cation, was observed in the ESCA spectrum on a time scale where no chemical equilibration process could have any effect. Subsequent ESCA studies by us (with Grunthaner's laboratory at Caltech's Jet Propulsion Laboratory) and by Dave Clark fully justified our previous results and conclusions. So did ever more advanced theoretical calculations.

Additional significant experimental studies were also carried out by others. Arnett reported valuable calorimetric studies. Saunders showed the absence of the deuterium isotopic perturbation of equilibrium ex-

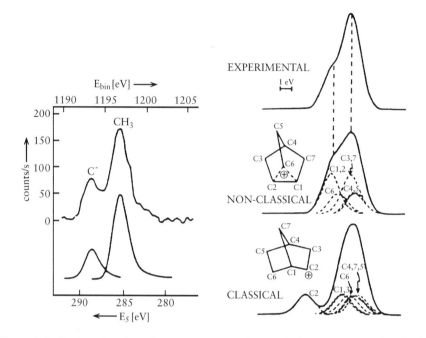

Figure 9.2. Carbon 1s photoelectron spectrum 1s core-hole-state spectra for the 2-norbornyl cation of *tert*-butyl cation and Clark's simulated spectra for the classical and nonclassical ions.

pected for a classical equilibrating system. Myhre and Yannoni, at extremely low (5 K) temperature, were able to obtain solid-state ^{13}C NMR spectra that still showed no indication of freezing out any equilibrating classical ions. The barrier at this temperature should be as low as 0.2 kcal/mol (the energy of a vibrational transition). Laube was able to carry out single crystal X-ray structural studies on substituted 2-norbornyl cations. Schleyer's theoretical studies including IGLO and related calculation of NMR shifts and their comparison with experimental data contributed further to the understanding of the σ-bridged carbonium ion nature of the 2-norbornyl cation. The classical 2-norbornyl cation was not even found to be a higher-lying intermediate!

As the norbornyl ion controversy evolved, it became a highly public and frequently very personal and bitter public debate. Saul Winstein suddenly died in the fall of 1969, shortly after the Salt Lake City symposium. To my regret, I seemed to have inherited his role in representing the bridged nonclassical ion concept in subsequent discussions. Although Brown frequently stated that he had not been given the opportunity to express his point of view in public debates, to my recollection, a number of these debates took place, including meetings at La Grand Motte (France), Bangor (UK), and Seattle, WA.

The 1983 Seattle American Chemical Society symposium was, in fact, the de facto "closing" of the long-running debate. Although in the heat of the debates some personal remarks were made on both sides (to my regret, as I expressed subsequently also), the experimental evidence at the time was so overwhelming that I concluded my presentation saying, "I don't intend to do any more research on the matter. There is nothing further to be discussed. . . ." My lecture was subsequently published (with Prakash and Saunders) under the title "Conclusion of the Classical-Nonclassical Ion Controversy Based on the Structural Study of the 2-Norbornyl Cation" in an article in the *Accounts of Chemical Research*. In the same issue, Brown wrote, "The nonclassical theory is not necessarily wrong, but it has been too readily accepted" (an obvious understatement concerning the enormous amount of work carried out on the topic). In any case, I kept my promise and have not done further work in the field. The chemical community, and later even the Swedish Academy, accepted the closure

Before our controversy in 1965 with Herbert Brown and Heinz Staab in Heidelberg

of the debate, mentioning inter alia in connection with my Nobel Prize the nonclassical norbornyl ion as one of the resolved carbocation structures.

This is the way the so-called nonclassical ion controversy ended. It basically centered on the question of whether the structure of carbocations, including rapidly equilibrating "classical" ions, can be depicted adequately by using only Lewis-type 2-electron 2-center covalent bonding or whether there are also bridged or σ-localized ions, whose structural depiction also necessitates 2-electron 3-center bonding, including higher-coordinate carbon. The result was a new understanding of the general bonding nature of carbon compounds as well as of the electron donor ability and reactivity of single bonds in saturated hydrocarbons and σ-bonded compounds in general.

Intensive, critical studies of a controversial topic also help to eliminate the possibility of errors. One of my favorite quotations is by George von Bekesy, a fellow Hungarian-born physicist who studied fundamental questions of the inner ear and hearing (Nobel Prize in medicine, 1961):

> "[One] way of dealing with errors is to have friends who are willing to spend the time necessary to carry out a critical examination of the experimental design beforehand and the results after the experiments have been completed. An even better way is to have an enemy. An enemy is willing to devote a vast amount of time and brain power to ferreting out errors both large and small, and this without any compensation. The trouble is that really capable enemies are scarce; most of them are only ordinary. Another trouble with enemies is that they sometimes develop into friends and lose a good deal of their zeal. It was in this way the writer lost his three best enemies. Everyone, not just scientists, need a few good enemies!"

Clearly there was no lack of devoted adversaries (perhaps a more proper term than enemies) on both sides of the norbornyl ion controversy. It is to their credit that we today probably know more about the structure of carbocations, such as the norbornyl cation, than about most other chemical species. Their efforts also resulted not only in rigorous studies but also in the development or improvement of many techniques. Although many believe that too much effort was expended on the "futile" norbornyl ion controversy, I believe that it eventually resulted in significant insights and consequences to chemistry. It affected in a fundamental way our understanding of the chemical bonding of electron-deficient carbon compounds, extending Kekulé's concept of the limiting ability of carbon to associate with no more than four other atoms of groups (see Chapter 10). An equally significant consequence of the norbornyl cation studies was my realization of the ability of saturated C-H and C-C single bonds to act as two-electron σ-donors towards strong electrophiles such as carbocations or other highly reactive reagents in superacidic systems, not only in intramolecular but also, as found subsequently, in intermolecular transformations and electrophilic reactions. The key for this reactivity lies in the ability to form two-electron three-center (2e-3c) bonds (familiar in boron and organometallic chemistry). The electrophilic chemistry of saturated hydrocarbons (including that of the parent methane) rapidly evolved based on the recognition of the concept and significance of hypercoordinated carbon, in short, hypercarbon, in chemistry (Chapter 10).

Once the direct observation of stable, long-lived carbocations generally in highly acidic (superacid) systems became possible, it led me to the recognition of the general concept of hydrocarbon cations, including the realization that five (and higher coordinate) carbocations are the key to electrophilic reactions at single bonds in saturated hydrocarbons (alkanes, cycloalkanes). in 1972, I offered a general definition of carbocations based on the realization that two distinct classes of carbocations exist (it seemed to be the logical name for all cations of carbon compounds, because the negative ions are called carbanions).

Trivalent {"classical"} carbenium ions contain an sp^2-hybridized electron-deficient carbon atom, which tends to be planar in the absence of constraining skeletal rigidity or steric interference. The carbenium carbon contains six valence electrons; thus it is highly electron deficient. The structure of trivalent carbocations can always be adequately described by using only two-electron two-center bonds (Lewis valence bond structures). CH_3^+ is the parent for trivalent ions.

Penta- (or higher) coordinate ("nonclassical"} carbonium ions contain five or (higher) coordinate carbon atoms. They cannot be described by two-electron two-center single bonds alone but also necessitate the use of two-electron three (or multi)-center bonding. The carbocation center always has eight valence electrons, but overall the carbonium ions are electron deficient because of the sharing of two electrons among three (or more) atoms. CH_5^+ can be considered the parent for carbonium ions.

Lewis' concept that a covalent chemical bond consists of a pair of electrons shared between two atoms is a cornerstone of structural chemistry. Chemists tend to brand compounds as anomalous whose structures cannot be depicted in terms of such valence bonds alone. Carbocations with too few electrons to allow a pair for each "bond" came to be referred to as "nonclassical," a name first used by Roberts for the cyclopropylcarbinyl cation and adapted by Winstein for the norbornyl cation. The name is still used, even though it is now recognized that, like other compounds, they adopt the structures appropriate for the number of electrons they contain with two-electron two- or two-electron three (even multi)-center bonding, not unlike the bonding principle established by Lipscomb (Nobel Prize, 1976) for boron compounds. The prefixes "classical" and "nonclassical," I believe, will gradually fade away as the general principles of bonding are recognized more widely.

Whereas the differentiation of trivalent carbenium and pentacoordinated carbonium ions serves a useful purpose in defining them as limiting cases, it should be clear that in carbocationic systems there always exist varying degrees of delocalization. This can involve participation by neighboring n-donor atoms, π-donor groups, or σ-donor C-H or C-C bonds.

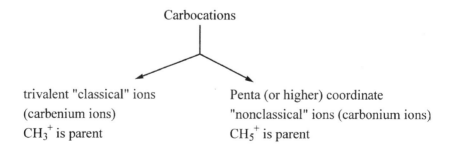

Carbocations

trivalent "classical" ions (carbenium ions) CH_3^+ is parent

Penta (or higher) coordinate "nonclassical" ions (carbonium ions) CH_5^+ is parent

Concerning carbocations, previous usage named the trivalent, planar ions of the CH_3^+ type *carbonium ions*. If the name is considered analogous to other *onium ions* (ammonium, sulfonium, phosphonium ions), then it should relate to the higher-valency or coordination-state carbocations. These, however, clearly are not the trivalent, but the penta- or higher-coordinated, cations of the CH_5^+ type. The earlier German and French literature, indeed, frequently used the "carbenium ion" naming for trivalent cations.

Trivalent carbenium ions are the key intermediates in electrophilic reactions of π-donor unsaturated hydrocarbons. At the same time, pentacoordinated carbonium ions are the key to electrophilic reactions of σ-donor saturated hydrocarbons through the ability of C-H or C-C single bonds to participate in carbonium ion formation.

Some characteristic bonding natures in typical nonclassical ions are the following.

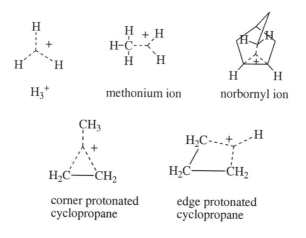

H₃⁺   methonium ion   norbornyl ion

corner protonated   edge protonated
cyclopropane   cyclopropane

Expansion of the carbon octet via *3d* orbital participation does not seem possible; there can be only eight valence electrons in the outer shell of carbon, a small first-row element. The valency of carbon cannot exceed four. Kekulé's concept of the tetravalence of carbon in bonding terms represents attachment of four atoms (or groups) involving 2e-2c Lewis-type bonding. However, nothing prevents carbon from also participating in multicenter bonding involving 2e-3c (or multicenter) bonds (see further discussion in Chapter 10).

Whereas the differentiation of limiting trivalent and penta- or higher-coordinate ions serves a useful purpose in establishing the significant differences between these ions, it must be emphasized that these represent only the extremes of a continuum and that there exists a continuum of charge delocalization comprising both intra- and intermolecular interactions.

Neighboring group participation (a term introduced by Winstein) with the vacant p-orbital of a carbenium ion center contributes to its stabilization via delocalization, which can involve atoms with unshared electron pairs (n-donors), π-electron systems (direct conjugate or allylic stabilization), bent σ-bonds (as in cyclopropylcarbinyl cations), and C-H and C-C σ-bonds (hyperconjugation).

Hyperconjugation is the overlap interaction of an appropriately oriented σ-bond with a carbocationic p-orbital to provide electron delocalization with minimal accompanying nuclear reorganization. Nuclear reorganization accompanying σ-bond delocalization can range from little or no rearrangement (hyperconjugation) to partial bridging involving some reorganization of nuclei (σ-participation) and to more extensive or complete bridging. Trivalent carbenium ions, with the exception of the parent CH_3^+, consequently always show varying degrees of delocalization. Eventually in the limiting case carbocations become pentacoordinated carbonium ions. The limiting cases define the extremes of the spectrum of carbocations.

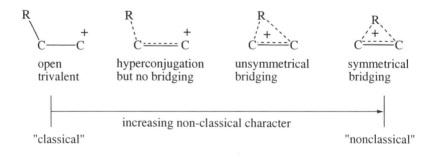

Under superacidic, low nucleophilicity so-called "stable ion conditions," developing electron-deficient carbocations do not find reactive external nucleophiles to react with; thus they stay persistent in solution stabilized by internal neighboring group interactions.

During my studies I realized that the formation of the σ-delocalized 2-norbornyl cation from 2-norbornyl precursors represented the equivalent of an intramolecular σ-alkylation where a covalent C_1-C_6 bond provided the electrons for the 2e-3c bonded bridged ion (by σ-participation).

In cases of more effective π-electron donor or n-donor neighboring groups, as is the case in forming β-phenylethyl (studied by Don Cram from UCLA; Nobel Prize in chemistry, 1987) or β-halogen bridged species, these have sufficient electrons to form 2e-2c bonds (with some intermediate delocalization).

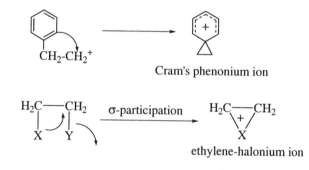

Cram's phenonium ion

ethylene-halonium ion

The intramolecular σ-delocalization in the norbornyl system aroused my interest in studying whether similar electrophilic interactions and reactions of C-H or C-C bonds are possible in intermolecular systems. This led to my discovery of the general electrophile reactivity of single bonds (Chapter 10). The long, drawn-out nonclassical norbornyl ion controversy thus led to an unexpected significant new chapter of chemistry. As frequently happens in science, the drive for understanding (for whatever reason) of what appear at the time to be rather isolated and even relatively unimportant problems can eventually lead to significant new concepts, new chemistry, and even practical applications. It justifies the need for exploration and study in the context of fundamental (basic) research even if initially no practical reasons or uses are indicated. The beauty of science lies in finding the unexpected, and, as

Niels Bohr was frequently quoted to have said, "you must be prepared for a surprise," but at the same time you must also understand what your findings mean and what they can be used for. To me, this is the lesson of the norbornyl ion controversy. I strongly believe it was not a waste of effort to pursue it, and eventually it greatly helped to advance chemistry to new areas of significance that are still emerging.

· 10 ·

From Kekulé's Four-Valent Carbon to Five- and Higher-Coordinate Hypercarbon Chemistry

One of the cornerstones of the chemistry of carbon compounds (organic chemistry) is Kekulé's concept, proposed in 1858, of the tetravalence of carbon. It was independently proposed in the same year by Couper who, however, got little recognition (vide infra). Kekulé realized that carbon can bind at the same time to not more than four other atoms or groups. It can, however, at the same time use one or more of its valences to form bonds to another carbon atom. In this way carbon can form chains or rings, as well as multiple-bonded compounds.

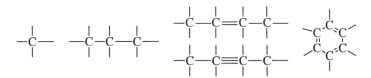

Kekulé claimed that the concept came to him during a late night ride on a London omnibus in 1854.

One fine summer evening, I was returning by the last omnibus, "outside" as usual, through the deserted streets of the metropolis, which are at other times so full of life. I fell into a reverie and lo! the atoms were gamboling before my eyes. . . . I saw how, frequently, two smaller atoms united to form a pair, how a larger one embraced two smaller ones; how still larger ones kept hold of three or even four of the smaller; whilst the whole kept whirling

in a giddy dance. I saw how the larger ones formed a chain. . . . I spent
part of the night putting on paper at least sketches of these dream forms.

—*August Kekulé, 1890.*

To what degree Kekulé's recollection was factual we don't know, but
Couper and Butlerov independently had developed similar, more well-
defined concepts of valence bonding, which may have not been entirely
unknown to Kekulé.

Another of Kelulé's revelations that supposedly came to him in a
dream was his famous structure of benzene. This related to how a
carbon chain can close into a ring. To satisfy the four valence of
carbon, this, of course, raised the need to involve alternating double
bonds.

I was sitting writing at my textbook, but the work did not progress; my
thoughts were elsewhere. I turned my chair to the fire, and dozed. Again
the atoms were gamboling before my eyes. This time the smaller groups
kept modestly in the background. My mental eye, rendered more acute by
repeated visions of this kind, could now distinguish larger structures
of manifold conformations; long rows, sometimes more closely fitted
together; all twisting and turning in snake-like motion. But look! What
was that? One of the snakes had seized hold of its own tail, and the
form whirled mockingly before my eyes. As if by a flash of lightning I
woke. . . . I spent the rest of the night working out the consequences of
the hypothesis. Let us learn to dream, gentlemen, and then perhaps we
shall learn the truth.

—*August Kekulé, 1865.*

In this case, we now know that the priority for the benzene structure
should at least be shared with the Austrian chemist Loschmidt, whose
book discussed much of the same concept but preceded Kekulé's.
Credit, however, for priority in science, as in other fields, is frequently
given to a large degree on the basis of how well the claim became
known, how widely it was communicated and disseminated. The lon-
gevity and staying power of the claimant also help. Couper's fundamental contribution, for example, is not widely recognized, perhaps
because he soon thereafter gave up science and never again pursued it.

Kekulé's fame and his extensive contributions to chemistry as a leading German professor of his time certainly overshadowed Loschmidt.

Dreams do not come complete with references and credit to preceding work by others. At the same time, it is also true that realizing the significance of a finding or observation (even if these were originated by someone else) and applying it to a broad concept of substantial significance and lasting importance is a genuine major contribution to science. Of course, I am not saying that proper credit should not always be given to preceding work or publications, but, regrettably, these sometimes tend to be lost over time. People eventually prefer to quote a single, easily available reference of a paper, book, or review, and even if these contain the original references, in subsequent quotations these are frequently not included. The "dominant" reference thus becomes the sole recognized source.

I certainly do not want to minimize Kekulé's major contributions to chemistry and their significance, but clearly there were—as is generally the case—other contributors who played a significant role and should be remembered.

Kekulé's four-valent carbon was explained later on basis of the atomic concept and the "rule of eight" valence electrons of the electronic theory of chemistry. From this, G. N. Lewis introduced the electron pair concept and that of covalent shared electron pair bonding (Lewis bond), which Langmuir (Nobel Prize in chemistry, 1932) further developed. It was Linus Pauling (Nobel Prize in chemistry, 1954), and others following him, who subsequently applied the principles of the developing quantum theory to the questions of chemical bonding. I prefer to use "chemical bonding" instead of "chemical bond," because, after all, in a strict sense the chemists' beloved electron pairs do not exist. Electrons move individually, and it is only the probability that they are found paired in close proximity that justifies the practical term of covalent electron-pair bonding. Pauling showed that electron pairs occupying properly oriented orbitals (which themselves are the preferred locations, but do not exist otherwise) result in the tetrahedral structure of methane (involving sp^3 hybridization). However, neither Lewis-Langmuir nor Pauling considered that an already shared electron pair could further bind an additional atom, not just two.

In the 1930s, Pauling still believed that diborane had an ethanelike structure and suggested this to Kharash during a visit to Chicago (recalled by H. C. Brown). It was Lipscomb (Nobel Prize, 1976), Pauling's student, who in the 1950s introduced the two-electron three-center (2e-3c) bonding concept into boron chemistry, also explaining the bridged structure of diborane.

It is remarkable that chemists long resisted making the connection between boron and electron-deficient carbon, which, after all, are analogs. I was thus given the opportunity to be able to establish the general concept of five and six coordination of electron-deficient carbon and to open up the field of what I called hypercarbon chemistry.

Organic chemists who are dealing with carbon compounds (or perhaps more correctly with hydrocarbons and their derivatives) have considered 2e-3c bonding limited to some "inorganic" or at best "organometallic" systems and have seen no relevance to their field. The long-drawn-out and sometimes highly personal nonclassical ion controversy was accordingly limited to the structural aspects of some, to most chemists rather obscure, carbocations. Herbert Brown, one of the major participants in the debate and, besides Lipscomb, one of the great boron chemists of our time, was steadfast in his crusade against bridged nonclassical ions. He repeatedly used the argument that if such ions existed, a new, yet unknown bonding concept would need to be discovered to explain them. This, however, is certainly not the case. The close relationship of electron-deficient carbocations with their neutral boron analogs has been frequently pointed out and discussed. Starting in a 1971 paper with DeMember and Commeyras, I pointed out the observed close spectral (IR and Raman) similarities between isoelectronic $^+C(CH_3)_3$ and $B(CH_3)_3$ and emphasized the point repeatedly thereafter. My colleagues Robert Williams, Surya Prakash, and Leslie Field did a fine job in carrying the carbocation, borane, and polyborane analogy much further and also reviewed the topic in depth in our book, *Hypercarbon Chemistry*.

On the basis of my extensive study of stable, persistent carbocations, reported in more than 300 publications, I was able to develop the general concept of carbocations referred to in Chapter 9. Accordingly, in higher-coordinate (hypercoordinate) carbonium ions, of which protonated methane CH_5^+ is the parent, besides two-electron two-center Lewis bonding, two-electron three-center bonding is involved.

Extensive ab initio calculations, including recent high-level studies, reconfirmed the preferred C_s symmetrical structure for the CH_5^+ cation, as we originally suggested with Klopman in 1969. The structure can be viewed as a proton inserted into one of the σ C-H bonds of methane to form a 2e-3c bond between carbon and two hydrogen atoms (or CH_3^+ binding H_2 through a long, weaker bonding interaction). At the same time, we already pointed out that ready bond-to-bond (isotopal) proton migration can take place through low barriers to equivalent or related structures that are energetically only slightly less favorable (which led more recently to Schleyer's suggestion of a fluxional, completely delocalized nature).

With Lammerstma and Simonetta in 1982, we studied the parent six-coordinate diprotonated methane (CH_6^{2+}), which has two 2e-3c bonding interactions in its minimum-energy structure (C_{2v}). On the basis of ab initio calculations, with Rasul we more recently found that the seven-coordinate triprotonated methane (CH_7^{3+}) is also an energy minimum and has three 2e-3c bonding interactions in its minimum-energy structure (C_{3v}). These results indicate the general importance of 2e-3c bonding in protonated alkanes.

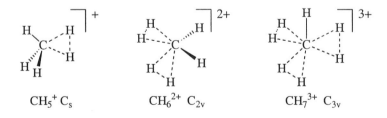

CH_5^+ C_s CH_6^{2+} C_{2v} CH_7^{3+} C_{3v}

Protonated methanes and their homologues and derivatives are experimentally indicated in superacidic chemistry by hydrogen-deuterium exchange experiments, as well as by core electron (ESCA) spectroscopy of their frozen matrixes. Some of their derivatives could even be isolated as crystalline compounds. In recent years, Schmidbaur has prepared gold complex analogs of CH_5^+ and CH_6^{2+} and determined their X-ray structures. The monopositively charged trigonal bipyramidal

$\{[(C_6H_5)_3PAu]_5C\}^+$ and the dipositively charged octahedral gold complex $\{[(C_6H_5)_3PAu]_6C\}^{2+}$ contain five- and six-coordinate carbon, respectively. Considering the isolobal relationship (i.e., similarity in bonding) between LAu^+ and H^+, the gold complexes represent the isolobal analogs of CH_5^+ and CH_6^{2+}.

The remarkable stability of the gold complexes is due to significant metal-metal bonding. However, their isolation and structural study are remarkable and greatly contributed to our knowledge of higher-coordinate carbocations.

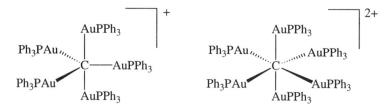

Boron and carbon are consecutive first-row elements. Trivalent carbocations are isoelectronic with the corresponding neutral trivalent boron compounds. Similarly, pentavalent monopositively charged carbonium ions are isoelectronic with the corresponding neutral pentavalent boron compounds. BH_5, which is isoelectronic with CH_5^+, has also C_s symmetrical structure based on high-level ab initio calculations. Experimentally, H-D exchange was observed in our work when BH_4^- was treated with deuterated strong acids, indicating the intermediacy of isopomeric BH_5. The first direct experimental observation (by infrared spectroscopy) of BH_5 has only recently been reported. The X-ray structure of the five-coordinate gold complex $[(Cy_3P^+)B(AuPPh_3)_4]$ was also reported by Schmidbaur. This square pyramidal compound represents the isolobal analog of BH_5, and further strengthens the relationship of the bonding nature in higher-coordinate boron and carbon compounds.

Similarly, as five- and six-coordinate CH_5^+ and CH_6^{2+} are isoelectronic with BH_5 and BH_6^+, respectively, seven-coordinate tripositively charged CH_7^{3+} is isoelectronic with the corresponding dipositively charged heptavalent boronium dication BH_7^{2+}. We have also searched for a minimum-energy structure of tetraprotonated methane, CH_8^{4+}.

However, CH_8^{4+} remains even computationally elusive because charge-charge repulsion appears to have reached its prohibitive limit. The isoelectronic boron analog BH_8^{3+}, however, was calculated to be an energy minimum.

As shown in studies of CH_5^+, CH_6^{2+}, and CH_7^{3+} and their analogs, carbon, despite its limiting tetravalence, can still bond simultaneously to five, six, or even seven atoms involving two-electron three (or multi)-center bonds. Such carbon atoms are called *hypercarbons* (short for hypercoordinated carbon atoms).

Because carbon is a first-row element unable to extend its valence shell, *hypervalence* cannot exist in carbon compounds, only *hypercoordination*.

Hypercarbon compounds contain one or more *hypercoordinated* carbon atoms bound not only by 2e-2c but also 2e-3c (or >3c) bonds.

The discovery of a significant number of hypercoordinate carbocations ("nonclassical" ions), initially based on solvolytic studies and subsequently as observable, stable ions in superacidic media as well as on theoretical calculations, showed that carbon hypercoordination is a general phenomenon in electron-deficient hydrocarbon systems. Some characteristic nonclassical carbocations are the following.

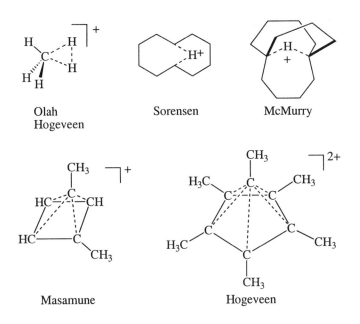

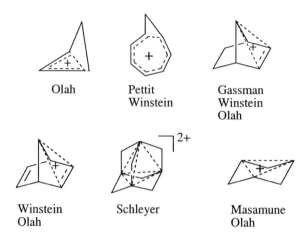

Olah	Pettit Winstein	Gassman Winstein Olah
Winstein Olah	Schleyer	Masamune Olah

According to early theoretical calculations Klopman and I carried out in 1971, the parent molecular ions of alkanes, such as CH_4^+, observed in mass spectrometry, also prefer a planar hypercarbon structure.

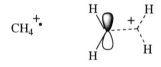

Even the CH_4^{2+} ion, as calculated later by Radom, has a similar planar C_{2v} structure.

Carbon can not only be involved in a single two-electron three-center bond formation but also in some carbodications simultaneously participate in two 2e-3c bonds. Diprotonated methane (CH_6^{2+}) and ethane ($C_2H_8^{2+}$), as well as the dimer of the methyl cation ($C_2H_6^{2+}$), are illustrative.

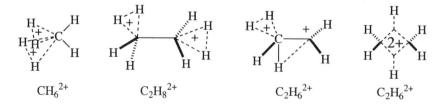

It was the study of hypercarbon-containing nonclassical carbocations that allowed us to firmly establish carbon's ability in a hydrocarbon system to bind simultaneously with five (or six or even seven) atoms or groups. It should be emphasized that carbocations represent only one class of hypercarbon compounds. A wide variety of neutral hypercarbon compounds, including alkyl (acyl)-bridged organometallics as well as carboranes, carbonyl, and carbide clusters, are known and have been studied. They are reviewed in my book *Hypercarbon Chemistry*. Representative examples are:

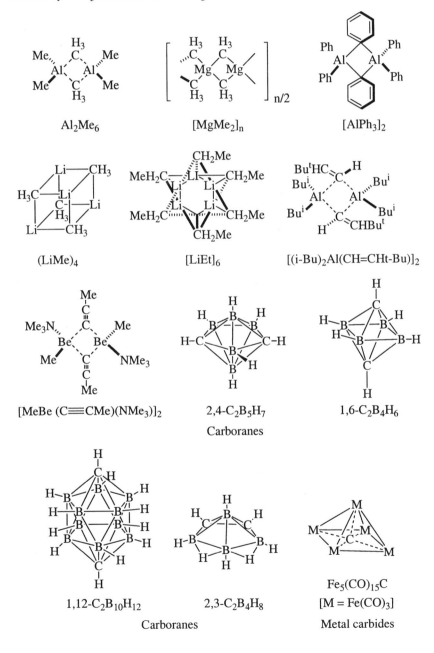

Al_2Me_6 $[MgMe_2]_n$ $[AlPh_3]_2$

$(LiMe)_4$ $[LiEt]_6$ $[(i\text{-}Bu)_2Al(CH=CHt\text{-}Bu)]_2$

$[MeBe\,(C{\equiv}CMe)(NMe_3)]_2$ $2,4\text{-}C_2B_5H_7$ $1,6\text{-}C_2B_4H_6$

       Carboranes

$1,12\text{-}C_2B_{10}H_{12}$ $2,3\text{-}C_2B_4H_8$ $Fe_5(CO)_{15}C$

               $[M = Fe(CO)_3]$

       Carboranes   Metal carbides

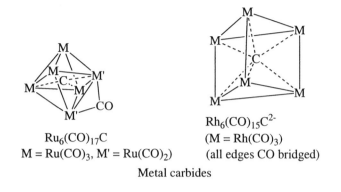

Ru$_6$(CO)$_{17}$C
M = Ru(CO)$_3$, M' = Ru(CO)$_2$)

Rh$_6$(CO)$_{15}$C^{2-}
(M = Rh(CO)$_3$)
(all edges CO bridged)

Metal carbides

The most studied hypercoordinate carbocation is the 2-norbornyl cation, around which the nonclassical ion controversy centered (Chapter 9).

The formation of the σ-delocalized norbornyl cation via ionization of 2-norbornyl precursors in low-nucleophilicity, superacidic media, as mentioned, can be considered an analog of an intramolecular Friedel-Crafts alkylation in a saturated system. Indeed, deprotonation gives nortricyclane.

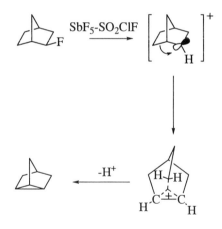

This realization led me to study related possible intermolecular electrophilic reactions of saturated hydrocarbons. Not only protolytic reactions but also a broad scope of reactions with varied electrophiles (alkylation, formylation, nitration, halogenation, oxygenation, etc.) were found to be feasible when using superacidic, low-nucleophilicity reaction conditions.

Protonation (and protolysis) of alkanes is readily achieved with superacids. The protonation of methane itself to CH_5^+, as discussed earlier, takes place readily.

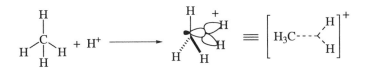

Acid-catalyzed isomerization reactions of alkanes as well as alkylation and condensation reactions are initiated by protolytic ionization. Available evidence indicates nonlinear but not necessarily triangular transition states.

$$R_3C\text{-}H + H^+ \longrightarrow [R_3C\text{---}H\text{---}H]^+ \longrightarrow R_3C^+ + H_2$$
$$\text{Linear}$$

$$R_3C\text{-}H + H^+ \longrightarrow \left[R_3C\text{---}\overset{H}{\underset{H}{\diagup}}\right]^+ \longrightarrow R_3C^+ + H_2$$
$$\text{Nonlinear}$$

The reverse reaction of the protolytic ionization of hydrocarbons to carbocations, that is, the reaction of trivalent carbocations with molecular hydrogen giving their parent hydrocarbons, involves the same five-coordinate carbonium ions.

$$R_3C^+ + \overset{H}{\underset{H}{|}} \rightleftharpoons \left[R_3C\text{----}\overset{H}{\underset{H}{\diagup}}\right]^+ \rightleftharpoons R_3C\text{-}H + H^+$$

The isomerization of butane to isobutane in superacids is illustrative of a protolytic isomerization, where no intermediate olefins are present in equilibrium with carbocations.

The superacid-catalyzed cracking of hydrocarbons (a significant practical application) involves not only formation of trivalent carbocationic sites leading to subsequent β-cleavage but also direct C-C bond protolysis.

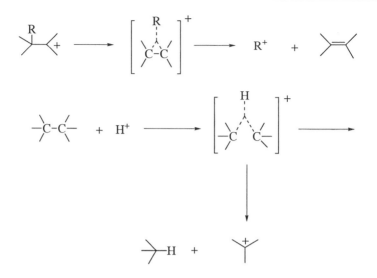

Whereas superacid (HF/BF$_3$, HF/SbF$_5$, HF/TaF$_5$·FSO$_3$H/SbF$_5$, etc.)-catalyzed hydrocarbon transformations were first explored in the liquid phase, subsequently, solid acid catalyst systems, such as those based on Nafion-H, longer-chain perfluorinated alkanesulfonic acids, fluorinated graphite intercalates, etc. were also developed and utilized for heterogeneous reactions. The strong acidic nature of zeolite catalysts was also successfully explored in cases such as H-ZSM-5 at high temperatures.

Not only protolytic reactions but also a whole range of varied electrophilic reactions can be carried out on alkanes under superacidic conditions.

$$E = D^+, H^+, R^+, NO_2^+, Hlg^+, HCO^+, etc.$$

Alkylation of isoalkanes with alkenes is of particular significance. The industrially used alkylation of isobutane with isobutylene to isooctane, is, however, de facto alkylation of the reactive isobutylene and not of the saturated hydrocarbon. Isobutane only acts as a hydride transfer agent and a source of the *tert*-butyl cation, formed via intermolecular hydride transfer. In contrast, when the *tert*-butyl cation is reacted with isobutane under superacidic conditions and thus in the absence of isobutylene, the major fast reaction is still hydride transfer, but a detectable amount of 2,2,3,3-tetramethylbutane, the σ-alkylation product, is also obtained. With sterically less crowded systems σ-alkylation becomes more predominant.

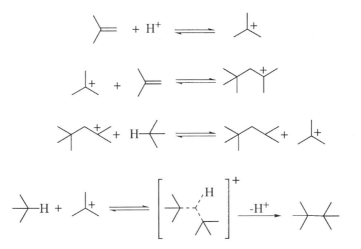

A fundamental difference exists between conventional acid-catalyzed and superacidic hydrocarbon chemistry. In the former, trivalent carbenium ions are always in equilibrium with olefins, which play the key role, whereas in the latter, hydrocarbon transformation can take place without the involvement of olefins through the intermediacy of five-coordinate carbocations.

The reaction of trivalent carbocations with carbon monoxide giving acyl cations is the key step in the well-known and industrially used Koch-Haaf reaction of preparing branched carboxylic acids from alkenes or alcohols. For example, in this way, isobutylene or *tert*-butyl alcohol is converted into pivalic acid. In contrast, based on the superacidic activation of electrophiles leading the superelectrophiles (see Chapter 12), we found it possible to formylate isoalkanes to aldehydes, which subsequently rearrange to their corresponding branched ketones.

These are effective high-octane gasoline additive oxygenates. The conversion of isobutane into isopropyl, methyl ketone, or isopentane into isobutyl, methyl ketone is illustrative. In this reaction, no branched carboxylic acids (Koch products) are formed.

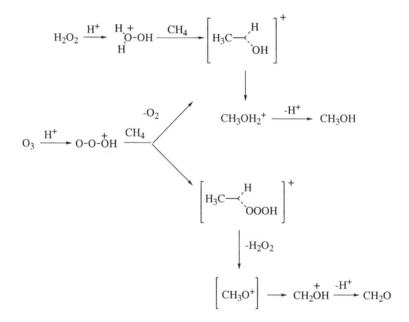

$$(CH_3)_3CH \xrightarrow[\text{HF/BF}_3]{CO} \left[(CH_3)_3C \overset{H}{\underset{CHO}{\diagdown}} \right]^+ \longrightarrow$$

$$(CH_3)_3CCHO \xrightarrow{HF-BF_3} (CH_3)_2CHCOCH_3$$

The superacid-catalyzed electrophile oxygenation of saturated hydrocarbons, including methane with hydrogen peroxide (via $H_3O_2^+$) or ozone (via HO_3^+), allowed the efficient preparation of oxygenated derivatives.

Because the protonation of ozone removes its dipolar nature, the electrophilic chemistry of HO_3^+, a very efficient oxygenating electrophile, has no relevance to conventional ozone chemistry. The superacid-catalyzed reaction of isobutane with ozone giving acetone and methyl alcohol, the aliphatic equivalent of the industrially significant Hock-reaction of cumene, is illustrative.

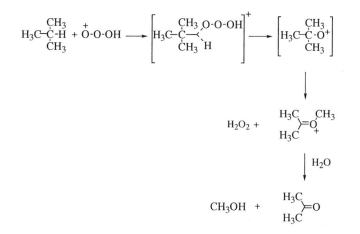

Electrophilic insertion reactions into C-H (and C-C) bonds under low-nucleophilicity superacidic conditions are not unique to alkane activation processes. The C-H (and C-C) bond activation by organometallic complexes, such as Bergman's iridium complexes and other transition metal systems (rhodium, osmium, rhenium, etc.), is based on somewhat similar electrophilic insertions. These reactions, however, cannot as yet be made catalytic, although future work may change this. A wide variety of further reactions of hydrocarbons with coordinatively unsaturated metal compounds and reagents involving hypercarbon intermediates (transition states) is also recognized, ranging from hydrometallations to Ziegler-Natta polymerization.

In the conclusion of my 1972 paper on the general concept of carbocations I wrote, "The realization of the electron donor ability of shared (bonding) electron pairs, including those of single bonds, should rank one day equal in importance with that of unshared (nonbonding) electron pairs recognized by Lewis. We can now not only explain the reactivity of saturated hydrocarbons and in general of single bonds in electrophilic reactions, but indeed use this understanding to explore new areas of carbocation chemistry."

This was one of the few times I ever made a prediction of the possible future significance of my chemistry. More than a quarter of a century later I take some satisfaction that I was correct and that, indeed, hypercarbon chemistry has a significant place on the wide palette of chemistry.

Carbon can extend its bonding from Kekulé's tetravalent limit to five-, and even higher-bonded (coordinate) hypercarbon systems. Higher-coordinate carbocations are now well recognized. They are the key to our understanding of the electrophilic reactivity of C-H or C-C single bonds and of hypercarbon chemistry in general. Some of their derivatives, such as related gold complexes, can even be isolated as stable crystalline compounds. The chemistry of higher-coordinate carbon (i.e., hypercarbon chemistry) is rapidly expanding, with many new vistas to be explored. The road from Kelulé's tetravalent carbon to hypercarbon chemistry took more than a century to travel. Carbon also unveiled other unexpected new aspects, for example, its recently discovered fullerene-type allotropes and their chemistry. There is no reason to believe that the new century just beginning will not bring further progress in the chemistry of carbon, including hypercarbon chemistry.

· 11 ·

The Nobel Prize:
Learning to Live with It and Not Rest on Laurels

Early in the morning on October 12, 1994, the phone rang in our home in Los Angeles on what promised to be another wonderful, sunny California day. Both my wife and I are early risers, and at 6 a.m. as usual I was already well into my morning routine, having breakfast and preparing for my day at the university. We had been away for a few days to the San Francisco area and had returned the previous afternoon, so I had a busy schedule ahead of me. The voice on the phone was that of the secretary of the Royal Swedish Academy, and the call changed my plans for the day and, in many ways, my relatively quiet, well-organized life. He informed me that the Academy had just voted to award me the 1994 Nobel Prize in chemistry and asked whether I would accept it (a question I believe recipients don't find difficult to answer). He then added his own congratulations and in good humor suggested that to establish the bona fide nature of his call and that he was not a prankster, he would put on some of my Swedish friends who were standing by and whose voices I probably would recognize. I indeed recognized them and was grateful for their good wishes.

My day, which had started out in its regular routine, from this point on became quite hectic. I barely had time to call my sons, who for some years at this time in October used to make "polite" and "understanding" remarks on the fact that their father was again not on the list of Nobelists (not that anyone really can or should expect this). I also called a few close friends, including my colleague Surya Prakash, as well as Katherine Loker and Carl Franklin, who had greatly helped my work since I moved to California and had always had great faith in me. From that point on our phone never stopped ringing. Amazingly, within a short while, even before I was able to shave and get fully dressed, some reporters started to show up at our house, which is on a remote cul-de-sac of a canyon on the top of Beverly Hills. By the time I finished my by then cold morning coffee I was giving the first of many interviews, to a reporter from the Voice of America, who was fastest in finding our home. One of the first questions he asked (to be repeated many times) was what I was going to do with the prize money. My spontaneous answer was that I was going to give the check to my wife, who handles our financial affairs. (More about this later on.)

The rest of the day was equally hectic. By the time Judy and I made it to the university, my colleagues and students had already organized a most touching welcome. I overheard my graduate student Eric Marinez proudly declaring "we won," reflecting the close-knit spirit of my group and perhaps the USC spirit of athletic competitiveness (which I must confess, however, had never enticed me). This was followed by a reception given by the president of the University, Steven Sample, always very supportive of my work, who made gracious remarks. Other events with much press and TV coverage followed. There was, understandably, substantial excitement on the USC campus because I was the first Nobel winner in the University's 114-year history. Attention on the campus was shifted at least for the time (but I hope also for the future) to academic achievements matching the long-standing recognition of USC as a premier athletic institution (with more Olympic medal winners than any other U.S. university and an outstanding record in football, baseball, track, and other sports).

Was I prepared for the Nobel Prize? I probably was, because over the years friends and colleagues had hinted that I had been nominated many times (of course, nobody really "knows" anything about the selection in advance). It is frequently said that the Nobel Prize represents the de facto end of the active research career of the recipients as they become public figures with little time left for scholarly work. However, I was determined that no prize or recognition would substantially change my life (see also Chapter 14). Furthermore, Judy, my life part-

The Olah group the day my Nobel Prize was announced with friends Bob Williams, Surya Prakash, Joe Casanova, and Reiko Choy, my secretary

ner and strength in all situations, never would have stood for changes affecting the real values we had come to hold and the life we had built for ourselves.

Living with the Nobel Prize in many ways is not easy, not least because it involves new obligations, public appearances, and speeches. However, it never brought any real problems to our life and relationship with our family, including my scientific family of more than 200 former students and postdocs, our friends, and colleagues. We did learn, nevertheless, to differentiate friends from acquaintances as well as public recognition (such as honorary degrees, memberships in academies, societies, etc.) dating before and after the Nobel Prize. The latter, I feel, often mean more interest in Alfred Nobel's prize than in the individual who won it. It is useful to keep this in mind before being

With Judy and sons George and Ron, and daughter-in-law Cindy at the USC Reception after the announcement of the prize, October 12, 1994

carried away by one's instantly acquired importance as well as the assumed universal wisdom and knowledge of all topics that Nobelists are frequently asked to comment on. It is clear that a physicist or a chemist cannot be an "expert" even in his broad and complex field of science, not to mention topics outside science. Becoming a universal oracle of wisdom on all matters is something I was never tempted in the slightest to pursue. It is perhaps a reflection on our times and society that publicity seems to be important regardless of the context. I certainly don't want to criticize others whose views are different from mine, but maintaining self-respect and a healthy appreciation of one's limitations and relative significance is very useful. I believe that I am reasonably well educated and well read in many areas outside my field, and I have certainly always had strong views. Not surprisingly, few, if any, gave attention to these views before my Nobel (except my long-suffering wife, who, over the years, must have heard some of them more often than she would like to remember). It was thus amusing to observe that after you receive the prize people start to believe that you are worth listening to.

I have many other recollections of the hectic days following the announcement of my prize. The flow of good wishes and personal messages was overwhelming. I made a firm commitment to answer them all in a personal way. We worked out an efficient system to do this, with the help of my wonderful secretary, Reiko Choy. I remember Bob Woodward many years ago took his secretary to Stockholm to answer good wishes directly from there, making it even more personal. I had not gone to this extreme, but answering properly and promptly was important to me, even more so because I had never gotten used to the fact that some people do no answer you or return your calls, not necessarily to disregard you, but perhaps because they are too busy or feel it is unimportant. In this regard, too, I can say that winning a Nobel Prize brings about changes. Your letters are always answered, and your phone calls are promptly returned.

Winning the Nobel Prize brings with it many invitations for various events. One of the first invitations my wife and I received was from President and Mrs. Clinton to come to the White House for a celebration of the year's American Nobel Prize winners. We had been to the White House before only as tourists guided through some of the public areas. As first-generation immigrants we certainly viewed the invitation as an honor, and we went to Washington with great anticipation. The event was indeed most impressive. It was also a pleasure to meet my fellow laureates, with whom we subsequently spent a memorable week in Stockholm. The President himself never showed up (we were told that he was very busy that day), but Mrs. Clinton and Vice President Gore were hosting the event. I was particularly impressed by Mrs. Clinton (although I must admit I don't share her politics). She spoke for 15 minutes quite impressively about the significance of science to the audience of some 150–200, including cabinet secretaries, members of Congress, directors of agencies, representatives of various scientific organizations, and invited various guests, without any notes or a teleprompter. The Vice President, in contrast, followed by reading a prepared text. He reflected on the impressive record of American Nobel winners. He mentioned that on the mantelpiece in one of the rooms of the White House is the Nobel medal of President Theodore Roosevelt, the first American Nobel Peace Prize winner (for settling the

Spanish War). On a lighter note, he apologized for the rather extensive security procedures involved for guests entering the White House. He recalled a story of the Swedish king who, while visiting Germany, was congratulated on having such a famous chemist as Scheele (one of the discoverers of oxygen) in his country. Neither the king nor anybody in his entourage had any idea who Scheele was. Nevertheless, the king gave an order to find him and, if he was a "decent fellow," to decorate him. Months later, it was reported to him that Scheele was found and, being a cavalry officer, was considered a reliable man and consequently honored. As it turned out, however, he was the wrong Scheele, having nothing to do with science or chemistry. Gore concluded his story that strict White House security also helps to prevent the admittance of the wrong scientists to events such as this.

Much has already been written about the Stockholm Nobel ceremonies. I would like, therefore, only to mention some personal recollections. The events, like all the activities of the Nobel Foundation, are

White House reception, November 1994, with Vice President Al Gore

superbly organized, with specific information covering all details provided in advance. The protocol assigns to each prize the privilege of
inviting ten guests to the major ceremonies and events. Because I won
the chemistry prize alone (prizes can be divided among no more than
3 persons), I could invite all ten guests myself (otherwise, like the prize
money, the invitations are also divided). We took our sons George and
Ron and our daughters-in-law Sally and Cindy with us and invited
Judy's cousin Steve Peto and his wife Magda, who live in Brussels. Two
close friends and colleagues, Surya Prakash and Peter Stang, also joined
us. Katherine Loker (a great benefactor whose name our Institute
bears) and Carl Franklin also accompanied us. Carl's late wife Caroline
used to tease me that if I ever got to Stockholm they would come along.
I am sure in spirit she was there beside Carl.

The Nobel Prizes are awarded each year on December 10, the anniversary of Alfred Nobel's death. The preceding week is taken up with
formal events and lectures by the laureates. The fairy tale week had
already started on our SAS flight to Stockholm, when the captain
greeted us on the public address system. At the airport there was a
reception, and then we were driven in a limousine, which was at our
disposal during our whole visit, to the famous Grand Hotel. We were
accompanied by our "attaché" Ms. Carina Martensson from the Swedish Foreign Office, who became our "guardian angel." She kept everything running smoothly and always got us to our many appointments
on time. She had attended the Monterey Institute of International Studies in California and was well acquainted with many California institutes, including Berkeley, Stanford, and Caltech (USC, however, must
have been a novelty to her). It is a tradition that the Swedish Foreign
Ministry assigns each laureate an attaché, which we learned is a prestigious and much sought after assignment, generally signaling a promising career. The Grand Hotel traditionally houses the Nobelists. In
earlier times when the events were much smaller, even the ceremonies
were held there. These have since moved to the Concert Hall and
Stockholm City Hall.

Alfred Nobel died in San Remo, Italy in 1896, and his will established the Prizes in physics, chemistry, physiology or medicine, and
literature and the Peace Prize. They were first given in 1901 and now

With our guardian angels in Stockholm (attache Carina Martensson with the briefcase)

span a century. In 1968, the Bank of Sweden established an additional Alfred Nobel Memorial Prize in Economic Sciences, which although not one of the original Nobel Prizes, is awarded with them and for all practical purposes has become the Nobel Prize in economics.

Whereas the events of the Nobel week are well reported, it is difficult to describe the spirit of all of Sweden during these events and their general meaning to the Scandinavian people. This struck me particularly during an event we attended honoring three high school science teachers selected for a special award for outstanding teaching. The event was televised all over Scandinavia and must have been most inspirational for students and their parents alike. In general, the importance of education and the significance of science for the future of mankind was emphasized throughout the week in a way that is unimaginable for Americans, who never see anything like this on their TV screens (and would never stand for it to replace their favorite shows and football games). Alfred Nobel's heritage thus has taken up national significance and acts as an inspiration for the whole country.

During our stay in Stockholm we attended the traditional Nobel concert in the famous Concert Hall (where later the awards themselves were presented). It was impressive to see the informality of the event. No particular security was evident for the Royal couple, who freely mingled with the invited guests and audience.

A luncheon for the American laureates and some 200 guests was given at the residence of Thomas Siebert, the American Ambassador to Sweden. The Ambassador, a former university classmate of President Clinton, and his wife were most gracious. We were also invited by the Hungarian Ambassador to a reception given for the two Hungarian-born winners (the economist John Harsanyi and myself). Little Hungary indeed can be proud to have produced such an impressive number of Nobel winners in proportion to its population (although nearly all emigrated and did most of their work abroad). Some believe that there must be some special talent in Hungarians for certain fields (besides the sciences, music, film making, engineering, and entrepreneurship are frequently mentioned). I believe, however, that the main reason was a good educational system, which is a more realistic explanation than is sometimes offered to explain the success of Hungarian-born scientists. During the Manhattan project, in which the Hungarian-born Neuman, Szilard, Teller, and Wigner played an important role, Enrico Fermi was quoted as suggesting that they were really visitors from Mars. They possessed advanced intelligence but found themselves in difficulty because they spoke English with a bad foreign accent, which would give them away. Therefore, they chose to pretend to be Hungarians, whose inability to speak any language but Hungarian without a thick foreign accent was well known (I am myself a good example of this). The story was memorialized in the book *Voice of the Martians* by the Hungarian physicist George Marx. It is an attractive saga, but in fact, the development of scientists depends to a great degree on a good education, which should always be emphasized.

Another event that stands out in my memory was the Nobel Lecture I gave before the Swedish Academy of Sciences, chaired by Professor Kerstin Fredga, its President and the leading Swedish space scientist. She is the daughter of the late Arne Fredga, a chemistry professor and long-time member of the Nobel Committee and the Nobel Foundation. I had known him and visited him in Uppsala years before; thus it was even more of a personal pleasure to meet his daughter. The only formal obligation connected with the Nobel Prize is to deliver this lecture in Stockholm. It is a distinct honor, and these lectures together with some biographical material appear in the annually published volumes *Les Prix Nobel.*

One evening during that week our friends Lars and Edith Ernster invited our family and some mutual friends for dinner. Lars (or Laci, as a born Hungarian) was a distinguished professor of biochemistry at Stockholm University who, regrettably, passed away recently. His wife Edith, a talented violinist, was the first female concert master of the Stockholm Opera Orchestra. I do not know how they managed the dinner for all of us in their apartment, but it was a most pleasurable and relaxing evening. The night before the award presentations we ourselves gave a dinner in a restaurant for our friends and guests, which also came off well.

Saturday, December 10, finally arrived. In the morning, there was a rehearsal of the award ceremony (with a stand-in for the King). Instead of the formal tails obligatory at the ceremony itself, I could wear a warm sweater and slacks, much more familiar and comfortable attire. We were instructed in detail and rehearsed the event. The award cer-

Nobel lecture, Stockholm, December 1994, with Professor Fredga, President, Royal Swedish Academy of Science

emony started on a dark, cold afternoon at exactly at 4:30 p.m. (punctuality, we learned, is obligatory in Sweden even for the King himself). With musical accompaniment and regal pomp, the ceremony proceeded with great precision. There is a time-honored sequence in which the prizes are presented, starting with physics, followed by chemistry, physiology or medicine, literature, and then economics (the peace prize is presented in Oslo on the same day).

When my turn came, I listened as Professor Salo Gronovitz, the Chairman of the Chemistry Nobel Committee, introduced me to the King and Queen and the formal assembly, including some former prize winners, members of the Swedish Academies, and visiting guests. He spoke initially in Swedish but ended with the following in English: "Professor Olah, I have in these few minutes tried to explain your immense impact on physical organic chemistry through your fundamental investigations of the structure, stability, and reactions of carbocations. In recognition of your important contribution, the Royal Swedish Academy of Sciences has decided to confer upon you this year's Nobel Prize for chemistry. It is an honor and pleasure for me to extend to you the congratulations of the Royal Swedish Academy of Sciences and to ask you to receive your Prize from the hands of His Majesty the King." With this, I stepped into the circle on the carpet marked with a large N and received my prize. The significance of the moment all at once struck me. The long journey that started in Budapest flashed through my mind and also some doubt about whether all this was real and was in fact happening. I forced myself, however, to the present and accepted my medal and the impressive diploma, which on one side has a water color painting, as I learned later, by the artist Bengt Landin.

After completion of the award ceremony in the Concert Hall we were driven to the city hall for the Nobel banquet. By this time it was a bitterly cold, windy evening. The courtyard of the city hall was lighted by the torches of hundreds of school children lining it. It was a most impressive sight, but we felt sorry for the children, who must have braved the weather for a long while.

The Nobel banquet in the Blue Hall of the Stockholm City Hall is traditionally attended, in addition to the Swedish Royal Family, by

Receiving the Nobel Prize from King Carl Gustav XVI, December 10, 1994

members of the Government and Parliament, diplomats, and some 1200 invited guests. Attendance at the banquet, we were told, is the most sought-after social invitation in Sweden.

We were first taken to a suite for cocktails and to organize for the entry into the banquet hall. The procession formed and then began. It was led by King Carl Gustaf XVI with Judy on his arm. This all followed protocol, according to which the King accompanies the wife of the physics winner, if there is a sole winner, or if this is not the case that of the chemistry winner, etc. Because the physics prize was shared that year, but I was the sole chemistry winner, Judy found herself leading the procession on the arm of the King. They were followed by the Queen, escorted by the President of the Nobel Foundation. The procession started on the mezzanine floor, which ran the length of the hall, and we marched down the impressive marble stairs into the large ground floor hall and then split to either side of a very long table. It was a most thrilling entrance, with trumpets blaring and everyone standing and cheering.

At the dinner, too, Judy was seated next to the King, and I was across the table at the side of the Queen. Queen Silvia is a beautiful and gracious lady. During the dinner I learned she was born in Heidelberg, Germany; her mother was from Brazil, where she spent part of her youth. We talked about many topics, not the least our children and her many duties, including the annual Nobel events, with which by now she was very familiar. The guests were seated at tables perpendicular to the main one, with our family and guests sharing a table with our Swedish chemist colleagues.

The banquet began with the Chairman of the Foundation proposing a toast to the King, who then answered with a silent toast to the memory of the great benefactor and philanthropist Alfred Nobel.

The banquet itself was a superbly choreographed event. Each course was brought in by hundreds of waiters descending the same staircase with musical and lighting accompaniment like a well-orchestrated ballet. The tables were beautifully set with gold-plated china and crystal specially designed and used only for the Nobel banquet and decorated with a profusion of flowers. These, I learned, were flown in from San Remo in Italy, where Alfred Nobel had a winter home (and where he died).

Toward the end of the dinner laureates representing each prize were called upon for a brief (not more than 3 minutes) remark. I generally don't write down any of my remarks or speeches, but this was a special occasion and I was requested to provide a text. Thus I can reprint here what I said:

> Your Majesties, Your Royal Highnesses, Ladies, and Gentlemen, I am most grateful for the honor bestowed on me today. Although receiving the Nobel Prize is the greatest satisfaction any scientist can experience, I consider it not only a personal acknowledgment, but also that of all my students, associates and colleagues whose dedicated work over the years allowed my field of chemistry, which is not frequently highlighted in public, to be recognized.

With Queen Silvia at the banquet dinner

There are many facets of chemistry. Mankind's drive to uncover the secrets of live processes and use of this knowledge led to spectacular advances in the biological and health sciences. Chemistry richly contributes to this by helping our understanding at the molecular level. Chemistry is, however, and always will be a central science of its own.

Chemists make compounds and strive to understand their reactions. My own interest lies in the chemistry of the compounds of the elements carbon and hydrogen, called hydrocarbons. These make up petroleum oil and natural gas and thus are in many ways essential for everyday life. They generate energy and heat our houses, fuel our cars and airplanes and are raw materials for most manmade materials ranging from plastics to pharmaceuticals. Many of the chemical reactions essential to hydrocarbons are catalyzed by acids and proceed through positive ion intermediates, called carbocations.

Remarks at the Nobel dinner

To be able to prepare and study these elusive species in stable form, acids billions of times stronger than concentrated sulfuric acid were needed (so called superacids). Some substituted carbocations, however, are remarkably stable and are even present in nature. You may be surprised to learn that the fine red wine we drank tonight contained carbocations which are responsible for the red color of this natural 12% or so alcoholic solution. I hope you enjoyed it as much as I did.

Chemistry does not always enjoy the best of reputations. Many of our plants and refineries are still potentially dangerous and may pollute their surroundings. At the same time our society enjoys a high standard of living not in small measure through the results of chemistry, which few would give up. I believe that chemistry can and will be able to bring about an equilibrium between mankind's needs and our environmental concerns. Chemistry will continue to benefit mankind in the spirit of Alfred Nobel, a fellow chemist whose example continues to inspire us all.

The audience particularly liked the red wine part. So did I, because usually it is not easy to explain what my "carbocations" are all about.

At the end of the banquet, we ascended the stairs to the Gold Room, where there was dancing till the early hours of the next day. Judy and I were, however, too exhausted after the long day, and after some private time with our Royal hosts we returned (or shall I say floated back) to our hotel.

The next day, Sunday the 11th, the Royal couple gave a dinner in their palace for the laureates and some 200 guests as a conclusion of the Nobel week. The austere, impressive palace I was told by the Queen is not suited to bringing up a family and they use it only for formal entertainment.

There was one other event for us to remember. While December 10 is Nobel day in Sweden, December 13 is St. Lucia's day. We were awakened early morning by a knock on our door and were greeted by a singing group of white-clad girls carrying candles and a traditional

With the King and Queen of Sweden after the Nobel banquet

breakfast to honor Santa Lucia. It was a moving and memorable morning.

During the week in Stockholm I also visited the University of Stockholm and the Royal Technological Institute. Subsequently, I also visited and lectured at Uppsala University and the Universities of Gothenburg and Lund. We were received everywhere with great friendship and hospitality, ending our most memorable trip in Copenhagen, from where we flew home. It was time to come back down to earth from the skies and resume our life.

The Nobel Prizes also come with a monetary award, which that year amounted to close to a million dollars. Besides paying taxes on it (the U.S. is the only country that taxes the Nobel Prizes), we donated part of it to help endow a chair in chemistry at USC as well as a chemistry prize in Hungary. The balance was shared with our children. It was thus not difficult to dispose of the prize money, but money, of course, is really not the essential part of the Nobel.

Concerning taxes, I was told that even the Nobel gold medal could be taxed based on its weight (seemingly the only value tax authorities put on it). I must confess, however, not to have paid duty on it, even after declaring it properly upon our return home. The nice lady customs inspector inquired as to how I acquired a gold medal. My wife told her that it was the Nobel medal I had just received in Stockholm. To my surprise, she not only knew what this was but shook my hand, saying that I was the second Nobel laureate she had had the pleasure to meet personally (the other was Linus Pauling). She decided that my "acquisition" of the medal did not come under the duty rules. I hope neither of us will get into any trouble for it.

Winning the Nobel Prize inevitably brings with it, besides a brief period of wider publicity (which in America evaporates particularly fast), a steady stream of invitations, varied honors and recognitions, as well as more general public involvement. Professors and scientists in American life are usually not exactly at the top of the "social ladder," nor are they used to much recognition. Personally, I rather like this, because it helps not to attach overgrown significance to one's importance, keeps one humanized, and, most important, allows one to stay centered without much distraction from one's work. It was, therefore, certainly an initial shock that for a while the limelight of the Prize seemed to intrude on my well-organized, quiet life. However, friends who have gone through this before assured me that this would soon slow down and that if you are set to continue your life in your "own way" it is possible to manage it. They were indeed right.

Being fairly strong willed, I was determined to continue my life with as few distractions as possible. Thus I was back at my university duties the day after our return from Stockholm. Research and teaching were always an integral part of my life, not just a "work habit," and I could not see any reason for change. I also did not feel ready for retirement, formal or de facto, while pretending otherwise. I like teaching and still feel able to contribute significantly to research. In short, I felt disinclined to rest on past laurels.

I must confess that it was easier to set my goals than to carry them out. All of us have only a finite amount of energy and available time. Furthermore, one of the most important aspects of my life always was to keep a balance between my professional and personal lives. Even so, I am afraid I had frequently shortchanged my family, particularly my children. Under no circumstances was I prepared to do this again, even more so as our adorable grandchildren, Peter and Kaitlyn, added new vistas and pleasures to our family life. On the other hand, I believe it is essential to stay active, and outside chemistry I do not know many ways to achieve this. However, I did make adjustments. I decided not to take on new graduate students, as the long-range major commitment needed to properly guide them through 4 or 5 years of research seemed unrealistic. I continued, however, to work closely with graduate students already in my group, helping and counseling students of the Loker Institute and working with my postdoctoral fellows, who increasingly became the mainstay of my research group. My close cooperation with my colleague and friend Surya Prakash also continued unabated in areas of mutual interest. He built up over the years an impressive research program on his own, while in areas of hydrocarbon chemistry and some other fields of mutual interest our joint research continues. I also extended my research into some exciting new areas (see Chapter 13).

One of the essentials of living with the Nobel Prize was to learn to firmly say *no*. Of the large number of invitations I receive for various events, recognitions, and lectures. I only accept a few. In any case, many of these are frequently not so much for me personally as for the lure of having a Nobel Prize winner for the event. Once you realize this, it is easier to decline. For a long time, I practically never traveled without Judy and do so even less now. We also travel less, although generally once a year we still go to Europe, never, however, if we can help it, in the winter. For us there is no place to live like Southern California. We swim each morning year round in our pool, to keep in shape and for our health, but also because we enjoy it so much. Where else could you do this, and afterwards go to your laboratory? We also decided to travel first class on long trips to lessen the stress associated with travel. A friend, Harry Gray, told us some time ago to do this, because if you do not, your children will. We agree and hope that our children understand.

As long as the enjoyment of teaching occasional courses, even to freshmen, doing research, and working with my younger associates and colleagues lasts and I feel able to make meaningful contributions, I intend to continue and not to rest on any past achievements. I believe this also keeps me active and interested past an age when others would have decided long ago to quit. Otherwise, continuing what I enjoy doing comes naturally, and I still have creative ideas and follow them up. One day when this is no longer the case, I fully intend to retire and start to act my age. While writing this I just learned that I am to receive the Arthur C. Cope Award of the American Chemical Society, which emphasizes achievements the significance of which became apparent in the past five years. This gives me particular pleasure because it acknowledges primarily my work done after my Nobel Prize. In any case, I try to follow the advice of a friend, Jay Kochi, who sent me a quotation from Edward Lavin's *Life Meditations*: "There are two things to aim at in life: the first, to get what you want, and after that, to enjoy it. Only the wisest of mankind achieve the second." I am not very wise, but I try to follow this advice.

· 12 ·

Post-Nobel Years:
From Superacids to Superelectrophiles

The Nobel Prize frequently puts so many new responsibilities and pre-occupations on the shoulders of the winners that it drastically changes their lives and affects their ability (or desire) to continue active research. I have heard some express the opinion that having received their prize at a more advanced age was in a sense a blessing, because it did not hinder their work throughout most of their career. This indeed may be the case, although Alfred Nobel's intention was to encourage and facilitate research through his prizes. However, as I mentioned in Chapter 11, I was, determined that the Prize should not affect my life, and certainly not my research, significantly. Now, six years later, and looking back on these post-Nobel years, I feel that I mostly succeeded. These were productive and in many ways most rewarding years of research. Helped by my dedicated younger colleagues and associates and by close collaboration with my colleague Surya Prakash, who increasingly took over many of the responsibilities and burdens associated with running our institute, I was able not only to continue my research but to extend it into new and challenging areas.

A significant part of my previous research was based on the study of carbocationic systems using superacids and their chemistry. The acids I was able to use and explore turned out to be many billions or even trillions of times stronger than previously recognized "strong" acids such as concentrated sulfuric acid. Acids are, in a general way, electron acceptors. Concerning protic (Brønsted) acids, in the condensed state there is no such thing as the unencumbered, naked proton (H^+). Having no electron, H^+ will always attach itself to any potential electron donors (whether n- [nonbonded], π-, or even σ-electron do-nors such as H_2). In the case of Lewis acids (such as $AlCl_3$, BF_3), their electron deficiency is affected by the stabilizing effect of ligands (neigh-boring groups) attached to the electron-deficient central atoms.

There are basically two approaches to enhancing the acidity of protic acids in the condensed state. The first involves decreasing the encum-brance of the proton by decreasing the nucleophilicity of the system. In aqueous media the proton is attached to water, forming the hydro-nium ion, H_3O^+ (in its hydrated forms). In anhydrous HF it is H_2F^+, in fluorosulfuric acid, $FSO_3H_2^+$, and so forth. The leveling effect means that no system can exceed the acidity of its conjugate acid. Thus no aqueous acid system can have acidity exceeding that of the conjugate acid of water, H_3O^+. Because the proton transfer ability of H_2F^+ is higher than that of H_3O^+ or $FSO_3H_2^+$, acid systems based on HF are stronger acids than those based on oxygenated acids.

The second approach is the use of increasingly lower-nucleophilicity anions and dispersion of the negative charge of the counter-anions (in the condensed state cations always must be balanced by anions). When anions are associated as, for example, via fluorine bridging going from SbF_6^- to $Sb_2F_{11}^-$, $Sb_3F_{16}^-$, etc., charge is increasingly dispersed. In $Al_2C_7^-$ etc. chlorine bridging also occurs but is generally less predom-inant in bringing about higher association than fluorine bridging.

The vastly increased acidity of superacidic systems resulted in the significant new field of superacid chemistry. I began to ask myself whether a similar but more general approach could be used to produce electrophiles of greatly enhanced electron deficiency and thus reactivity. Over the years, there were a number of unexpected results in my own research work, as well as some previously unexplained observations buried in the literature, that seemed worth pursuing.

Because of the mentioned leveling effect of the solvent (or excess acid itself acting as such) the acidity cannot exceed that of its conjugate acid. In the case of water the limiting acidity is that of H_3O^+. Proton-ated water, H_3O^+ (hydronium ion), was first postulated in 1907, and its preeminent role in acid-catalyzed reactions in aqueous media was first realized in the acid-base theory of Brønsted and Lowry. Direct experimental evidence for the hydronium ion in solution and in the solid state came eventually from IR, Raman, and neutron diffraction studies. The gaseous ion was observed in mass spectrometric studies, including its clustered forms with up to some 20 water molecules, H_3O^+ $(H_2O)_n$ (Kabarle). This ion-neutral association (clustering) is somewhat like solvation in solution. In superacid solution, the hydronium ion was studied by ^1H and ^{17}O NMR spectroscopy. Eventually, hydronium ion salts with a variety of counter-ions such as SbF_6^-, AsF_6^-, and BF_4^- were isolated and their X-ray structure was obtained (Christe). All the experimental and high-level ab initio theoretical data support a pyramidal geometry for the hydronium ion.

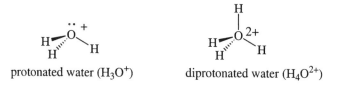

protonated water (H_3O^+)   diprotonated water (H_4O^{2+})

Isotopomeric hydronium ions such as H_2DO^+ and HD_2O^+ were also prepared in $HSO_3F{:}SbF_5$-D_2O/SO_2ClF solutions and characterized by ^1H and ^2H NMR spectroscopy, showing no exchange. Subsequently, however, Prakash and I, to our initial surprise, found that in the even stronger superacidic systems such as $HF/DF{:}SbF_5$, with increasing amounts of SbF_5 hydrogen/deuterium exchange takes place. Because no deprotonation equilibrium of the hydronium ion occurs in the extremely strong superacidic media and because H/D exchange occurs only with increasing acidity (i.e., with $HF/DF{:}SbF_5$ but not with $HSO_3F/DSO_3F{:}SbF_5$), the results suggested that the isotopic exchange takes place via diprotonated (deuteriated) isotopomeric H_4O^{2+} involved. However, in a limited equilibrium of mono- with diprotonated water, under extremely strongly acidic conditions the nonbonded electron pair of the oxygen atom of H_3O^+, despite being a positive ion, undergoes a second protonation.

$$HD_2O^+ \xrightleftharpoons{\ H^+\ } H_2D_2O^{2+} \xrightleftharpoons{\ -D^+\ } H_2DO^+$$

$$\Big\Updownarrow H^+$$

$$H_3O^+ \xrightleftharpoons{\ -D^+\ } H_3DO^{2+}$$

To probe the structure and stability of the H_4O^{2+} dication, ab initio theoretical calculations have also been carried out. The H_4O^{2+} dication is isoelectronic with H_4N^+, CH_4, and BH_4^-, and has a tetrahedral minimum structure. Apparently, the double positive charge can be adequately accommodated in the tetrahedral structure to prevent spontaneous fragmentation. It should be mentioned that in H_3O^+ the oxygen is indeed highly negatively charged, the positive charge being essentially on the hydrogen atoms. Whereas thermodynamically H_4O^{2+} is unstable toward deprotonation to H_3O^+, it has significant kinetic barrier to deprotonation.

It must be recognized that the calculational data refer to idealized dilute gas-phase and not to condensed-state conditions, where solvation or clustering may have a stabilizing influence. In particular in small dications, solvation tends to diminish the charge-charge repulsion effect and thus to bring H_4O^{2+} into a thermodynamically more accessible region. The extra proton of H_4O^{2+} thus might be shared by more than one H_3O^+ in a dynamic fashion.

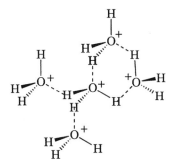

In subsequent studies Schmidbaur succeeded in preparing and isolating the stable BF_4^- salt of $^{2+}O[AuP(Ph_3)]_4$ and determined its X-ray structure. Although, clearly, charge is significantly delocalized in the heavy metal ligand belt, nevertheless the isolation of an $^{2+}OL_4$ species is remarkable.

The significance of the possible diprotonation of water under extremely acidic conditions directly affects the question of acid strength achievable in superacidic systems. The leveling effect mentioned above limits the acidity of any system to that of its conjugate acid. Thus, in aqueous systems, the acidity cannot exceed that of H_3O^+ present in equilibrium with H_2O.

$$H_2O + H^+ \rightleftharpoons H_3O^+$$

In superacidic systems, water is completely protonated and no equilibrium containing free water is indicated. However, the nonbonded electron pair of H_3O^+ is still a potential electron donor and at very high acidities can be further protonated (however limited the equilibrium with H_3O^+ may be). Thus the acidity of such superacidic systems can exceed that of H_3O^+ and the "leveling out" is by that of H_4O^{2+}. We found that similar situations exist with other electrophiles, raising their electrophilic nature (electrophilicity) substantially.

$$H_2O + H^+ \not\rightleftharpoons H_3O^+ + H^+ \rightleftharpoons H_4O^{2+}$$

Alkyloxonium ions are derived from the parent hydronium ion, H_3O^+, by substituting one, two, or all three hydrogen atoms with alkyl groups (i.e., ROH_2^+, R_2OH^+, R_3O^+). Meerwein prepared salts of the trimethyl and triethyl oxonium ions, R_3O^+, and found them to be excellent alkylating agents for various heteroatom nucleophiles but not for alkylating hydrocarbons. In conjunction with protic superacids, such as FSO_3H, CF_3SO_3H, or $FSO_3H:SbF_5$ (magic acid), however, we found that they readily alkylate aromatics. For example, benzene and toluene were methylated and ethylated with trimethyl- and triethyloxonium salts in the presence of superacids. The protolytic activation of trialkyloxonium ions points to protonation (protosolvation) of the nonbonded electron pair of their oxygen atom (similar to that in H_3O^+), which enhances the electrophilicity of the alkyl groups. Lewis acid complexation has a similar effect.

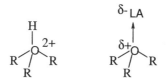

Acid-catalyzed alkylation of aromatics with alcohols themselves is widely used. Whereas tertiary (and secondary) alcohols react with relative ease, following a carbocationic mechanism, alkylation with primary alcohols such as methyl alcohol is achieved only in case of more reactive aromatics and under forcing conditions. Superacidic activation, however, again allows ready methylation of aromatics.

Similar activation takes place in the carbonylation of dimethyl ether to methyl acetate in superacidic solution. Whereas acetic acid and acetates are made nearly exclusively using Wilkinson's rhodium catalyst, a sensitive system necessitating carefully controlled conditions and use of large amounts of the expensive rhodium triphenylphosphine complex, ready superacidic carbonylation of dimethyl ether has significant advantages.

$$CH_3OCH_3 + CO \xrightarrow{\text{"}H^+\text{"}} CH_3COOCH_3$$
$$\downarrow H_2O$$
$$CH_3OH + CH_3CO_2H$$

Acyl cations are relatively weak electrophiles. This is easily understood, because their structure is of a predominantly linear carboxonium ion nature, with the neighboring oxygen atom delocalizing charge and limiting their carbocationic nature.

$$\overset{+}{R}-C\overset{..}{=}\overset{..}{O}: \quad \longleftrightarrow \quad R-C\overset{+}{\equiv}\overset{..}{O}:$$

In their reactions with suitable nucleophiles, such as π-aromatics or heteroatom donor nucleophiles, the readily polarizable linear acylium ions shift a π-electron pair to oxygen, bending the ions and developing an empty p-orbital at the carbocationic center. This enables the reaction with aromatics. The acetylation of benzene can be depicted as

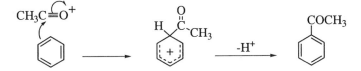

The lack of reactivity of acyl cations such as the acetyl cation with deactivated aromatics or saturated hydrocarbons is therefore not unexpected. However, under conditions of superacidic activation the oxygen atom of acetyl cation is protosolvated, greatly decreasing neighboring oxygen participation, enhancing the carbocationic nature and thus acylating ability. This explains, for example, why the acetyl cation in superacidic media reacts with isobutane via hydride transfer involving a protoacetyl dication-like species, but in weaker acidic or aprotic media there is no reaction.

$$H_3C-C\overset{+}{\equiv}O: \ + \ HF \ + \ BF_3 \ \rightleftharpoons \ CH_3\overset{+}{C}\overset{\overset{+}{O}H}{\diagup}$$

$$\Big| (CH_3)_3CH$$

$$\left[(H_3C)_3C-\overset{H}{\underset{\underset{OH}{||}}{\diagup}} CH_3 \right]^{2+}$$

$$CH_3CH=\overset{+}{O}H \ + \ (CH_3)_3C^+$$

Apart from Brønsted acid activation, the acetyl cation (and other acyl ions) can also be activated by Lewis acids. Although the 1:1 CH_3COX-AlX_3 Friedel-Crafts complex is inactive for the isomerization of alkanes, a system with two (or more) equivalents of AlX_3 was found by Volpin to be extremely reactive, also bringing about other electrophilic reactions.

$$C_nH_{2n+2} \ \xrightarrow[20\,°C,\ 30\ min]{RCOX +2\ AlCl_3} \ i\text{-}C_4H_{10} \ + \ i\text{-}C_5H_{12} \ +$$

$$i\text{-}C_4H_9COR \ + \ \left[\overset{}{\diagup}C-C\overset{}{\diagdown} \right]_m$$

Volpin called the $CH_3COCl\cdot 2AlCl_3$ complex an aprotic superacid. The results indicate that the acetyl cation is activated by further O-complexation with a second molecule of AlX_3 (the equivalent of protonation or protosolvation).

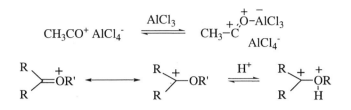

The results of activation of acyl cations led to our study of other *carboxonium* ions. Carboxonium ions are highly stabilized compared to alkyl cations. As their name indicates, they have both carbocationic and oxonium ion nature.

Although the latter predominates, strong neighboring oxygen participation (in Winstein's terms) delocalizes charge heavily onto oxygen, and renders the ions substantially oxonium ion-like. Superacidic activation protonates (protosolvates) the oxygen atoms, thus decreasing their participation, and enhances carbocationic nature and reactivity.

Carboxonium ions, for example, do not react with alkanes. However, in superacid solution acetaldehyde (or acetone), for example, readily reacts with isobutane involving diprotonated, highly reactive carbocationic species.

The acid-catalyzed isomerization of alkanes (of substantial practical interest in production of high octane gasoline) is a consequence of the ready rearrangement of the involved alkyl cations (Chapter 10). Carbonyl compounds (aldehydes, ketones) contain only a mildly electron-deficient carbonyl carbon atom of the polarized carbonyl group, which is insufficient to bring about rearrangements. Under superacidic activation, as we found in our studies, the carbonyl oxygen atoms not only protonate but are further protosolvated, significantly decreasing neighboring oxygen participation and development of carbocationic character, which allows rearrangements to occur. An example is the rearrangement of pivaldehyde to methyl, isopropyl ketone.

Acetic acid and other carboxylic acids are protonated in superacids to form stable carboxonium ions at low temperatures. Cleavage to related acyl cations is observed (by NMR) upon raising the temperature of the solutions. In excess superacids a diprotonation equilibrium, indicated by theoretical calculations, can play a role in the ionization process.

$$CH_3CO_2H \xrightleftharpoons{H^+} CH_3CO_2H_2^+ \xrightleftharpoons{H^+} CH_3CO_2H_3^{2+}$$

$$\downarrow \quad -H_3O^+ \;\Big|\; H^+$$

$$CH_3CO^+ \xrightleftharpoons{H^+} CH_3\overset{+}{C}=\overset{+}{O}H$$

Protonation of formic acid similarly leads, after the formation at low temperature of the parent carboxonium ion, to the formyl cation. The persistent formyl cation was observed by high-pressure NMR only recently (Horvath and Gladysz). An equilibrium with diprotonated carbon monoxide causing rapid exchange can be involved, which also explains the observed high reactivity of carbon monoxide in superacidic media. Not only aromatic but also saturated hydrocarbons (such as isoalkanes and adamantanes) can be readily formylated.

$$HCO_2H \xrightleftharpoons{H^+} HCO_2H_2^+ \xrightleftharpoons{H^+} HCO_2H_3^{2+}$$

$$\downarrow \quad -H_3O^+ \;\Big|\; H^+$$

$$CO \xrightleftharpoons{H^+} HCO^+ \xrightleftharpoons{H^+} H\overset{+}{C}\overset{+}{O}H$$

$$H_2O + CO_2 \xrightleftharpoons{\quad\quad} H_2CO_3$$

The equilibrium of carbon dioxide with water forming carbonic acid has long been recognized and is of great significance. Although carbonates are among the most abundant minerals on earth, free carbonic acid is elusive in solution. Protonated carbonic acid in superacidic media is, however, remarkably stable. We observed years ago (Chapter 7) that by dissolving carbonates or hydrogen carbonates in cold FSO_3H-SbF_5 (or other superacids) no CO_2 evolution occurred. The solutions decompose only at around 0°C to give hydronium ion and carbon dioxide. The SbF_6^- salt of $H_3CO_3^+$ could recently be isolated and its X-ray structure determined. The close analogy between protonated carbonic acid and the guanidinium $[C(NH_2)_3^+]$ ion explains its stability. Both are highly resonance stabilized. The guanidinium ion undergoes further protonation in superacid solution, and we observed the dication in such media by NMR spectroscopy. Similarly protonated carbonic acids can also readily undergo a second protonation to $H_4CO_3^{2+}$. The protolytic cleavage of carbonic acid, which leads to carbon dioxide, should proceed through protonated and diprotonated CO_2, i.e., $O=C\text{-}OH^+$ and $HO\text{-}COH^{2+}$.

$$CO_3^{2-}/ \atop HCO_3^- \xrightarrow[\text{-80°C}]{\text{FSO}_3\text{H-SbF}_5/\text{SO}_2} C(OH)_3^+ \xrightarrow{\text{-10 --> 0°C}} CO_2 + H_3O^+$$

$$C(OH)_3^+ \underset{}{\overset{H^+}{\rightleftharpoons}} (HO)_2\overset{+}{\underset{}{C}}\overset{+}{OH}_2 \xrightarrow[\text{-H}_3\text{O}^+]{H^+} CO_2H_2^{2+}$$

$$\Big\downarrow \text{-H}^+$$

$$CO_2 \xleftarrow{H^+} CO_2H^+$$

Protonated and diprotonated carbonic acid and carbon dioxide may also have implications in biological carboxylation processes. Although behavior in highly acidic solvent systems cannot be extrapolated to in vivo conditions, related multidentate interactions at enzymatic sites are possible.

Similar to oxonium ions, our studies of *sulfonium* ions also showed protosolvolytic activation in superacids to give sulfur superelectrophiles. The parent sulfonium ion (H_3S^+), for example, gives H_4S^{2+} (diprotonated hydrogen sulfide) in superacids.

$$H_2S \xrightarrow{H^+} H_3S^+ \underset{}{\overset{H^+}{\rightleftharpoons}} H_4S^{2+}$$

Various sulfonium and carbosulfonium ions show remarkably enhanced reactivity upon superelectrophilic activation, similar to their oxygen analogs; so do selenonium and telluronium ions. The alkylating ability of their trialkyl salts, for example, is greatly increased by protosolvation.

Halonium ions, including hydrido or alkylhalonium ions, are similarly protolytically activated, indicative of protonation of the nonbonded electron pairs of their halogen atoms.

$$HBr \xrightarrow{H^+} H_2Br^+ \overset{H^+}{\rightleftharpoons} H_3Br^{2+}$$

$$CH_3\overset{+}{X}CH_3 \overset{H^+}{\rightleftharpoons} CH_3\underset{H}{\overset{+}{X}}CH_3{}^{2+}$$

Various other heteroatom-substituted carbocations were also found to be activated by superacids. α-Nitro and α-cyanocarbenium ions, $R_2C^+NO_2$ or R_2C^+CN, for example, undergo O- or N-protonation, respectively, to dicationic species, decreasing neighboring nitrogen participation, which greatly enhances the electrophilicity of their carbocationic centers.

$$R_2\overset{+}{C}-NO_2 \overset{H^+}{\rightleftharpoons} R_2\overset{+}{C}-\overset{+}{N}O_2H$$

$$R_2\overset{+}{C}-CN \overset{H^+}{\rightleftharpoons} R_2\overset{+}{C}-\overset{+}{C}NH$$

To explain the experimentally observed high reactivity of HCN and alkyl nitriles under superacidic condition, Shudo found that in the Gatterman and Houben-Hoesch reactions, diprotonated HCN (or nitriles) are involved as the de facto reagents ($HC^+N^+H_2$, $RC^+N^+H_2$).

$$RCN \overset{H^+}{\rightleftharpoons} R-\overset{+}{C}=NH$$

$$\downarrow$$

$$R\overset{+}{C}=\overset{+}{N}H_2 \overset{H^+}{\rightleftharpoons} R-C\equiv\overset{+}{N}H$$

An efficient nitrating agent, which I introduced starting in my early research in Hungary, is nitronium fluoroborate $NO_2{}^+BF_4{}^-$ and related nitronium salts (Chapter 7). They are generally used in aprotic solvents (nitromethane, sulfolane, dichloromethane). Whereas a large variety of aromatics are nitrated with them in excellent yields, strongly deactivated aromatics such as *meta*-dinitrobenzene are not further nitrated under these conditions. In contrast, in superacidic FSO_3H or CF_3SO_3H solution, nitronium salts nitrate *meta*-dinitrobenzene to 1,3,5-trinitrobenzene in 70% yield as well as other deactivated nitrofluorobenzenes.

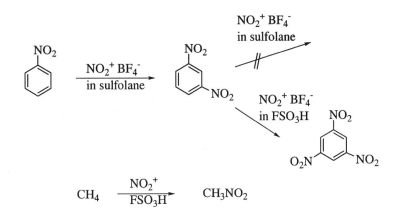

Similarly, whereas nitronium salts show no reactivity toward methane in aprotic media, nitration takes place, albeit in low yield, in FSO₃H to give nitromethane. These data indicated that the nitronium is activated by the superacidic media. Its highly increased reactivity is explained by the finding that the linear nitronium ion, despite being a positively charged species, undergoes protonation (protosolvation) to the protonitronium dication (NO_2H^{2+}), an extremely reactive superelectrophile. The linear nitronium $O=N^+=O$ itself has neither a vacant atomic orbital on nitrogen nor a low-lying molecular orbital. Its electrophilic nature is mainly due to its polarizability when interacting with π-donor aromatic nucleophiles. In nitrating aromatics, when NO_2^+ approaches the π-donor substrate the latter polarizes NO_2^+ by displacing an N-O bonding electron pair onto oxygen. This causes bending and the development of an empty p-orbital on nitrogen, allowing eventual formation of a nitrogen-carbon bond.

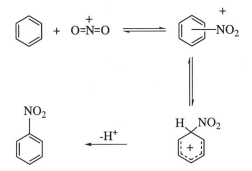

The finding that highly deactivated aromatics do not react with NO_2^+ salts is in accord with the finding that their greatly diminished π-donor ability no longer suffices to polarize NO_2^+. Similarly, σ-donor hydrocarbons such as methane (CH_4) are not able to affect such polarization. Instead, the linear nitronium ion is activated by the superacid. Despite the fact that NO_2^+ is a small, triatomic cation, the non-bonded electron pairs on oxygen can further interact with the acid. This interaction can be considered to be protosolvation (the electrophilic equivalent of usual nucleophilic solvation), which tends to bend the linear NO_2^+ ion. In the limiting case, de facto protonation forms NO_2H^{2+}. The protionitronium (which has even been observed by Helmut Schwarz by mass spectrometry) is a highly reactive "superelectrophile." Protolytic solvation, however, can suffice in itself to achieve activation without necessarily forming the free NO_2H^{2+} dication.

$$O\!=\!\overset{+}{N}\!=\!O \quad \overset{H^+}{\rightleftharpoons} \quad \left[O\!=\!N\overset{O}{\underset{O}{\big\langle}}_{OH} \right]^{2+}$$

Whereas the proton (H^+) can be considered the ultimate Brønsted acid (having no electron), the helium dication (He^{2+}) is an even stronger, doubly electron-deficient electron acceptor. In a theoretical, calculational study we found that the helionitronium trication (NO_2He^{3+}) has a minimum structure isoelectronic and isostructural with NO_2H^{2+}.

Charge-charge repulsion effects in protolytically activating charged electrophiles certainly play a significant role, which must be overcome. Despite these effects multidentate protolytic interactions with superacids can take place, increasing the electrophilic nature of varied reagents.

We found a way to overcome charge-charge repulsion when activating the nitronium ion when Lewis acids were used instead of strong Brønsted acids. The Friedel-Crafts nitration of deactivated aromatics and some aliphatic hydrocarbons was efficiently carried out with the $NO_2Cl/3AlCl_3$ system. In this case, the nitronium ion is coordinated to $AlCl_3$.

$$\left[O{=}N{\diagup}^{O}{\longrightarrow} AlCl_3 \right]^{+} \quad Al_2Cl_7^{-}$$

The protolytic activation of hydrocarbon ions (carbocations) can also be achieved. We have found that the electrophilic nature of trivalent carbenium ion centers can be greatly enhanced by decreasing neighboring group participation not only by π- or n-donor ligands but also by hyperconjugatively interacting σ-donor alkyl groups. For example, the hyperconjugative effect of the methyl groups in the *tert*-butyl cation can be protosolvolytically decreased, resulting in the limiting case in a dipositive carbenium, carbonium dication.

$$(CH_3)_3C^+ + H^+ \longrightarrow (CH_3)_2\overset{+}{C}\overset{+}{C}H_4$$

This may be a factor while acid-catalyzed transformations (isomerization, alkylation) of saturated hydrocarbons proceed preferentially in excess strong acid media.

My group's studies on superelectrophilic activation found that not only superacidic solutions but also solid acid systems can bring about such activation. Solid strong acids, possessing both Brønsted and Lewis acid sites, are of increasing significance. They range from supported or intercalated systems to highly acidic perfluorinated resinsulfonic acids (such as Nafion-H and its analogues) and to certain zeolites (such as H-ZSM-5) and acidic oxides. One of the major difficulties in characterizing solid acids is the accurate determination of their acidity. Frequently used methods are based on kinetic rather than thermodynamic measurements, which can give data on catalytic activity of the solid acids but not necessarily on their acidities.

To explain how solid acids such as Nafion-H or HZSM-5 can show remarkable catalytic activity in hydrocarbon transformations, the nature of activation at the acidic sites of such solid acids must be considered. Nafion-H contains acidic $-SO_3H$ groups in clustered pockets. In the acidic zeolite H-ZSM-5 the active Brønsted and Lewis acid sites are in close proximity ($\sim$2.5 Å).

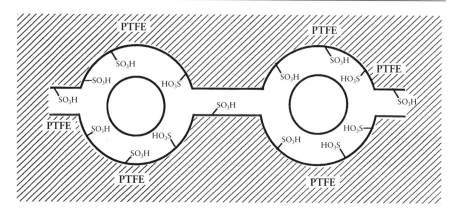

Figure 12.1. Perfluorinated resinsulfonic acid similar to Nafion-H, showing clustering of SO₃H groups; PTFE = poly(tetrafluoroethylene).

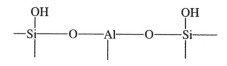

Brønsted and Lewis acid sites in zeolites

In these (and other) solid superacid catalyst systems, bi- or multidentate interactions are thus possible, forming highly reactive intermediates. This amounts to the solid-state equivalent of protosolvation resulting in superelectrophilic activation.

Nature is able to perform its own transformations in ways that chemists have only begun to understand. At enzymatic sites many significant transformations take place that, in a generalized sense, are also acid catalyzed (for example, electron-deficient metal ion-catalyzed processes). Because the unique geometry at enzyme sites can make multidentate interactions possible, these superelectrophilic chemical activations may have their equivalent in some enzymatic process. In recent years Thauer, for example, found an active metal free dehydrogenase enzyme. Because the involved mechanism cannot be that of metal coordination, it was suggested to involve carbocationic transformations (in a way again similar to superelectrophilic activation). Although I was never involved with enzymatic studies, sometimes in the literature reference is made to this "Olah-type" enzyme.

My recent work on superelectrophiles emerged from my previous studies on superacidic carbocationic and onium systems. It led to the realization that a variety of electrophiles capable of further interaction (coordination) with strong Brønsted or Lewis acids can be greatly activated by them. Examples mentioned were onium and carboxonium ions, acyl cations, halonium, azonium, carbazonium ions, and even certain substituted carbocations and the like. This activation produces what I suggested should be called *superelectrophiles*, that is, electrophiles of doubly electron-deficient (dipositive) nature whose reactivity significantly exceeds that of their parents. Superelectrophiles are the de facto *reactive intermediates* of many electrophilic reactions in superacidic systems (including those involving solid superacids) and should be differentiated from energetically lower-lying, much *more stable intermediates*, which frequently are observable and even isolable but are not necessarily reactive enough without further activation.

Examples of some superelectrophiles so far studied and their parents are

It should be recognized that superelectrophilic reactions can also proceed with only "electrophilic assistance" (solvation, association) by the superacids without forming distinct depositive intermediates. Protosolvolytic activation of electrophiles should always be considered in this context.

· 13 ·

Societal and Environmental Challenges of Hydrocarbons:
Direct Methane Conversion, Methanol Fuel Cell, and Chemical Recycling of Carbon Dioxide

Another area of my post-Nobel research that turned into a major continuing effort evolved from the realization that our hydrocarbon resources, the marvelous gift of nature in the form of petroleum oil and natural gas, are finite and not renewable.

The rapidly growing world population, which was 1.6 billion at the beginning of the twentieth century, has now reached 6 billion. Even if mankind increasingly exercises population control, by mid-century we will reach around 9.5–10 billion. This inevitably will put enormous pressure on our resources, not the least on our energy resources. For its survival, mankind needs not only food, clean water, shelter, and clothing, but also energy. Since the cave man first managed to light fire, our early ancestors burned wood and subsequently other natural sources. The industrial revolution was fueled by coal, and the twentieth century added oil and gas and introduced atomic energy.

World Population (in millions)

1650	1750	1800	1850	1900	1920	1952	2000
545	728	906	1,171	1,608	1,813	2,409	6,000

When fossil fuels such as coal, oil, or natural gas (i.e., hydrocarbons) are burned in power plants to generate electricity or to heat our homes and fuel our cars and airplanes, they form carbon dioxide and water. Thus they are used up and are nonrenewable (at least on the human time scale). We must find ways to replace our diminishing natural resources with hydrocarbons made by ourselves in a renewable, economical, and environmentally adaptable, clean way. In my view, this represents a most significant challenge for chemistry in the twenty-first century.

In an increasingly technological society, the world's per capita resources have difficulty keeping up. Society's demands, however, must be satisfied while at the same time safeguarding the environment to allow future generations to continue to enjoy planet Earth as a hospitable home. Establishing an equilibrium between mankind's needs and the environment is a challenge we must meet.

Nature has given us a remarkable gift in the form of oil and natural gas. However, what was created over the ages, man is using up rather rapidly. The large-scale use of petroleum and natural gas to generate energy, and also as raw materials for diverse man-made materials and products (fuels, plastics, pharmaceuticals, dyes, etc.), developed mainly during the twentieth century. The U.S. energy consumption (and that of the majority of the world) is very heavily based on fossil fuels. Atomic energy and other sources (hydro, geothermal, solar) represent only 11–12%.

U.S. Energy Sources (%)

Power Source	1960	1970	1990
Oil	48	46	41
Natural gas	26	26	24
Coal	19	19	23
Nuclear energy	3	5	8
Hydro, geothermal, wind, solar, etc. energy	4	4	4

Some Western industrial countries, in contrast, get a significant part of their energy from nonfossil sources, such as hydro and atomic energy (vide infra). Regrettably, alternate energy sources such as hydro-energy or even solar energy (via various conversion process) still represent only a small part of our overall energy picture. Whereas atomic energy is man's best hope for clean, practically unlimited energy for the foreseeable future, we must make it safer and solve problems of disposal and storage of radioactive waste byproduct.

Power Generated in Industrial Countries by Nonfossil Fuels (1990)

	Nonfossil Fuel Power (%)		
Country	Hydroenergy	Nuclear Energy	Total
France	12	75	87
Canada	58	16	74
Former West Germany	4	34	38
Japan	11	26	37
U.K.	1	23	24
Italy	16	0	16
U.S.	4	8	12

Oil use has grown to the point where the world consumption is 70–75 million barrels a day (a barrel equals 42 gallons, i.e., some 160 liters) or some 10 million metric tons. Oil and gas are mixtures of hydrocarbons. As already mentioned, once we burn hydrocarbons, they are irreversibly used up and are not renewable on the human time scale. Fortunately, we still have significant worldwide fossil fuel reserves, including heavy oils, shale and tar-sands, and even larger deposits of coals (a complex mixture of carbon compounds more deficient in hydrogen) that can be eventually utilized, albeit at a higher cost. I am not suggesting that our resources will run out in the foreseeable future, but it is clear that they will become scarcer and much more expensive and will not last for very long. With the world population at 6 billion and rapidly growing, the demand for oil and gas will inevitably increase. It is true, however, that in the past dire predictions of fast-disappearing oil and gas reserves were always incorrect, and today our proven reserves are significantly higher than they were decades ago.

Proven oil reserves, instead of being depleted, as a matter of fact, tripled over the last 30 years and now are a trillion barrels. Natural gas reserves have grown even more. This seems so impressive that most people assume that there can be no oil or gas shortage in sight. However, inevitably increasing consumption by a growing world population

Recognized Oil and Natural Gas Reserves (in billion tons) from 1960 to 1990

Year	Oil	Natural Gas
1960	43.0	15.3
1965	50.0	22.4
1970	77.7	33.3
1975	87.4	55.0
1980	90.6	69.8
1986	95.2	86.9
1987	121.2	91.4
1988	123.8	95.2
1989	136.8	96.2
1990	136.5	107.5

and the drive for improving standards of living makes it more realistic to consider per capita reserves. Doing this, it becomes evident that our known reserves, if we go on using them as in the past, will last for half a century. Even if we consider all other factors (new findings, savings, alternate sources, etc.) in the twenty-first century, we will face a major problem. Oil and gas will not be exhausted overnight, but market forces of supply and demand will start to drive prices up to levels that no one wants even to contemplate now ($100/barrel for oil may be just a beginning). By the second half of the century, if we do not find new solutions, mankind will face a real crisis with grave economic and societal consequences.

All of mankind wants the advantages an industrialized society can give its citizens. We all rely on energy, but the level of consumption is vastly different in the industrialized versus the developing world. For example, oil consumption per capita in China currently is 5 barrels/year, but it is well over tenfold that amount in the United States. China's oil use may, even under conservative estimates, double in a decade, whereas the bulk of its energy needs will continue to come from coal. This increase alone equals the amount of U.S. consumption, reminding us of the size of the problem we are facing. This estimate does not consider that vast numbers of Chinese (or Indians, etc.) would, instead of riding bicycles, drive their own cars and use other energy-consuming conveniences to the level common in the industrial world. However, they will (and may already) ride mopeds and enjoy other technical advances of our time, necessitating increasing use of energy. Do we in the industrialized world have a monopoly on a better life? I certainly don't think so.

Generating energy by burning nonrenewable fossil fuels including oil, gas and coal at the present level, as I pointed out, will be possible only for a relatively short term, and even this causes serious environmental problems (vide infra). The advent of the atomic age opened up for mankind wonderful new possibilities but also created dangers and concerns for safety. I believe that it is tragic that these considerations, however justified, practically brought further development of atomic energy to a standstill in most of the Western world. Whether we like it or not, in the long run we have no known alternative to relying increasingly on atomic energy, but we must solve safety and environmental problems. Finding solutions is essential and entirely within the capability of mankind if we have the will and determination for it. After all, during World War II we created the atom bomb by a great national effort. Could the same country that harnessed the energy of the atom in a wartime effort not also solve problems for its safe, peaceful use? I believe we can and will.

We continue to burn our hydrocarbon resources mainly to generate energy and to use them as fuels. Diminishing supplies (and sharply increasing prices) will lead inevitably to the need to supplement or make them ourselves by synthetic manufacturing. Synthetic oil or gasoline products will be, however, much costlier. Nature's petroleum oil and natural gas are still the greatest bargains we will ever have. A barrel of oil sells for around $30 (with market fluctuation). No synthetic manufacturing process will be able to come even close to this price, and we will need to get used to this, not as a matter of any government policy, but as a fact of market forces over which we have little control.

Synthetic oil is feasible and can be produced from coal or natural gas via synthesis gas (a mixture of carbon monoxide and hydrogen obtained from incomplete combustion of coal or natural gas). However, these are themselves nonrenewable resources. Coal conversion was used in Germany during World War II by hydrogenation or, mainly, by syn-gas (CO-H_2)-based Fischer-Tropsch synthesis. In South Africa during the 1960–1970 boycott years similar, improved technology was used. The size of these operations, however, amounted to barely 0.3% of present U.S. consumption. The Fischer-Tropsch synthesis is also highly energy-consuming, wasting half of the coal or natural gas used. Complex hydrocarbon mixtures are obtained that are difficult to refine, and overall, it hardly can be the technology of the future. New and more economical processes are needed. Some of the needed new basic chemistry and technology is now evolving, and a major research effort of our Loker Institute is directed to this goal.

We still have significant natural gas resources. They may even significantly expand not only from new discoveries (countless exploration efforts are going on and gas resources are increasingly used from all areas of the world, from polar regions to the depths of the oceans) but also from such unconventional sources as vast deposits of solid methane hydrates (clathrates of methane with surrounding water molecules) found under the tundras of Siberia and other polar regions but also on the continental shelves in the oceanic depths. Besides microbial conversion of algae or biomass, one way such deposits of methane could be formed is by reduction of carbon dioxide dissolved in the water of the oceans by hydrogen sulfide vented from the depth of the earth crest through fissions, with the help of microorganisms and without using the sun's energy (which cannot penetrate to these depths).

There is also the promise of finding large amounts of "deep" methane formed not from biomass but by some abiological processes from carbonates or even carbides formed from carbon-containing asteroids that hit the earth over the ages under the harsh prebiological conditions of our planet.

We should also utilize liquid hydrocarbons, which frequently accompany natural gas. These so-called "natural gas liquids" currently have little use besides their caloric heat value. They consist mainly of saturated straight hydrocarbons chains containing 3–6 carbon atoms, as well as some aromatics. As we found (Chapter 8), it is possible by superacidic catalytic treatment to upgrade these liquids to high-octane, commercially usable gasoline. Their use will not per se solve our long-range need to find new ways to produce hydrocarbons when our natural resources are becoming depleted, but in the short run they could be valuable resources.

Most natural gas (i.e., methane) is still burned to produce energy. However, methane (CH_4) should be recognized as a most valuable source for higher hydrocarbons, because in it nature provides us with the highest possible (4:1) hydrogen-to-carbon ratio. The question is, how can we utilize methane to obtain higher hydrocarbons and their derivatives from it directly, without wastefully burning it (albeit incompletely) first to synthesize gas (CO and H_2) to be used in Fischer-Tropsch chemistry?

Extensive studies were carried out in recent years to find ways for the selective oxidative conversion of methane to higher hydrocarbons. Because combining two methane molecules to form ethane and hydrogen is itself exothermic by some 16 kcal/mol, oxidative removal of H_2 is needed to make the reaction feasible.

$$2\,CH_4 \longrightarrow C_2H_6 + H_2 + 16\,kcal/mol$$

Some selective reactions were found, including oxidative coupling to ethane, but these reactions usually gave only low yields. In contrast, higher-yield reactions generally give low selectivities. Superacidic oxidative condensation of methane to higher hydrocarbon was explored in our work using varied metal halide-containing superacids. However, these are themselves used up as oxidizing agents to remove hydrogen. More effective catalytic oxidative coupling conditions must be found for practical processes.

A new approach we found is based on the initial bromination of methane to methyl bromide, which can be effected with good selectivity, although still in relatively low yields. Methyl bromide is easily separated from excess methane, which is readily recycled. Hydrolysis of methyl bromide to methyl alcohol and its dehydration to dimethyl ether are readily achieved. Importantly, HBr formed as by product can be oxidatively recycled into bromine, making the overall process catalytic in bromine.

$$CH_4 + Br_2 \xrightarrow{\text{cat.}} HBr + CH_3Br$$

$$\xrightarrow{H_2O} CH_3OH + HBr$$

$$2\,HBr \xrightarrow{1/2\,O_2} H_2O + Br_2$$

$$CH_4 + 1/2\,O_2 \longrightarrow CH_3OH$$

Although many problems still remain to be overcome to make the process practical (not the least of which is the question of the corrosive nature of aqueous HBr and the minimization of formation of any higher brominated methanes), the selective conversion of methane to methyl alcohol without going through syn-gas has promise. Furthermore, the process could be operated in relatively low-capital-demanding plants (in contrast to syn-gas production) and in practically any location, making transportation of natural gas from less accessible locations in the form of convenient liquid methyl alcohol possible.

The question arises: What is the real significance of being able to convert natural gas (i.e., methane) directly into methyl alcohol? The answer is that methyl alcohol can be subsequently converted using either zeolitic catalysts (Mobil's HZSM-5 or UOP's varied zeolites) or by our nonzeolitic, bifunctional acid-base catalytic chemistry (using catalysts such as WO_3/Al_2O_3) to the gasoline range or aromatic hydrocarbons. Furthermore, it is also possible by using related catalysts to convert methyl alcohol or dimethyl ether to ethylene or propylene, respectively.

$$2\,CH_3OH \longrightarrow CH_2{=}CH_2 + 2\,H_2O$$

The petrochemical industry knows how to run processes to convert ethylene and propylene (obtained from petroleum fractions by hydrocracking or from saturated hydrocarbons by dehydrogenation) to practically any hydrocarbon product or derivative. If diminishing petroleum oil reserves put increasing pressure on their availability and drive prices up steeply, methyl alcohol could become a key raw material for hydrocarbons. Of course, as long as syn-gas is used to prepare methyl alcohol, its energy-demanding production (wasting half of the natural gas or coal resources) would not solve the problem. Therefore, to be able to produce methyl alcohol directly from methane (natural gas) without going through syn-gas or from carbon dioxide (vide infra) is of great importance.

Another even more significant use of methyl alcohol can be as a fuel in its own right in fuel cells. In recent years, in cooperation with Caltech's Jet Propulsion Laboratory (JPL), we have developed an efficient new type of fuel cell that uses methyl alcohol directly to produce electricity without the need to first catalytically convert it to produce hydrogen.

The fuel cell is a device that converts chemical energy directly into electrical energy. The fuel cell was discovered in 1839 by William Grove, who found that the electrolysis of water in dilute sulfuric acid using platinum electrodes can be reversed. When water is electrolyzed, hydrogen and oxygen are formed. What Grove found was that if hydrogen and oxygen are combined in a cell over platinum electrodes in dilute sulfuric acid, water is formed and, simultaneously, electricity is produced. On the anode the fuel is oxidized; on the cathode oxygen is reduced. Grove's discovery, however, remained a curiosity. Only a century later was work on the hydrogen-oxygen fuel cell taken up again. Attempts were made unsuccessfully to use cheaper metal catalysts. After WW II alkalies, and later phosphoric acid, were used instead of sulfuric acid to ensure good contact between the gas, the electrolyte, and the solid electrode. These types of fuel cells slowly started to receive limited application in static installations. With the advent of the space age, the need for improved fuel cells arose. For the Gemini and Apollo space programs, and later for the space shuttle, JPL developed and built oxygen-hydrogen fuel cells using pressurized liquefied gases. These cells worked well, but the handling of liquefied hydrogen and oxygen is not only cumbersome but also highly dangerous and can result in explosions. The need for a versatile, relatively light-weight and simple fuel cell for transportation and other uses, replacing inefficient heavy batteries, resulted in our cooperation in the 1990s with colleagues from JPL and Caltech in developing a new generation of fuel cell. The cell we developed is no longer based on hydrogen [which also can be produced from various liquid fuels by a catalytic converter device (called reformer), i.e., a small syn-gas-producing unit, with CO separated (and oxidized to CO_2) from H_2]. Instead, our fuel cell directly utilizes methanol, a convenient, liquid fuel, which is oxidized at the anode. The proton exchange membrane (PEM) initially used was of perfluorinated ionomer polymer (DuPont's Nafion), the electrodes being Ph-Ru. The membrane contains large amounts of absorbed water to conduct protons efficiently; thus liquid water is present. However, because not only protons but methanol itself can cross through the membrane, diffusion of methanol from the anode to the cathode, became a major problem. To overcome the crossover we eventually succeeded in developing an efficient new membrane, which eliminated the problem.

The schematic diagram of the liquid-feed direct methanol fuel cell (DMFC) is shown in Figure 13.1.

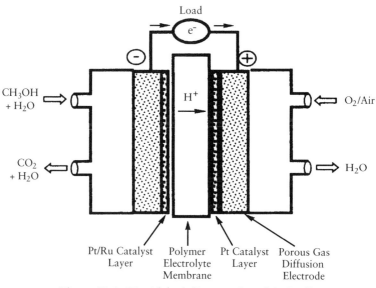

Figure 13.1. Liquid-feed direct methanol fuel cell.

The chemical reactions involved are:

Anode $CH_3OH + H_2O \longrightarrow CO_2 + 6H^+ + 6e^-$

Cathode $1.5\,O_2 + 6\,H^+ + 6\,e^- \longrightarrow 3\,H_2O$

Overall $CH_3OH + 1.5\,O_2 \longrightarrow CO_2 + 2\,H_2O$

In the overall fuel cell process methanol is oxidized to produce carbon dioxide and water while producing electricity. Unlike processes that burn fossil fuels to liberate heat, the fuel cell converts chemical energy directly to electrical energy, and therefore it is not limited in efficiency by the so-called Carnot principle (according to which the efficiency of any reversible heat engine depends only on the temperature range through which it works rather than on the properties of the fuel). Because the working temperature cannot be increased excessively, heat engines have limited efficiencies. The efficiencies of power plants (using, for example, a combination of gas and stream turbines) can be enhanced to 40%, but efficiencies of internal combustion engines are less than 20%. In contrast, the DMFC approaches 40% efficiency, and this can improve further.

Because the DMFC does not use hydrogen, it is not only a greatly simplified system but also a safe and convenient one. Methanol is a water-soluble liquid and can be safely transported and dispensed by existing infrastructure (at gasoline stations). The high efficiency of the fuel cell over the internal combustion engine, I believe, can make methanol the transportation fuel of the future. Its use can be also equally important in electricity generation not only in power plants but as a replacement for polluting and less efficient diesel generators in many parts of the world where the electric grid does not yet exist (or as a reliable emergency electricity source). It should also be useful in smaller, portable applications ranging from motor scooters to cellular phones, computers, and other electric devices, replacing batteries that have limited capacity and lifetime. In a battery, the chemicals accounting for the electricity-producing process are enclosed. Once they are used up, the battery must be recharged or replaced. In a fuel cell, external fuel is passed through continuously. Thus the lifetime basically depends only on the fuel, such as methanol. Our fuel cell has operated for periods in excess of 5000 hours.

Burning of any hydrocarbon (fossil fuel) or, for that matter, any organic material converts its carbon content to carbon dioxide and its hydrogen to water. Because power plants and other industries emit large amounts of carbon dioxide, they contribute to the so-called greenhouse warming effect on our planet, which causes significant en-

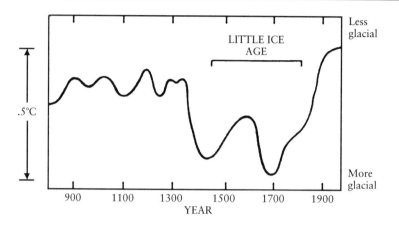

Figure 13.2. The change in temperature in Europe over the past 1000 years.

vironmental concerns. This was first indicated in 1898 in a paper by the Swedish chemist Svente Arrhenius (Nobel Prize 1903). The warming trend of our earth can be evaluated only over longer time periods, but, from available data, there is a relationship between increasing CO_2 content of the atmosphere and the temperature (Figs. 13.2 and 13.3).

The increase of carbon dioxide in the atmosphere (even if its overall amount is only 0.035%) affects our global climate, although other

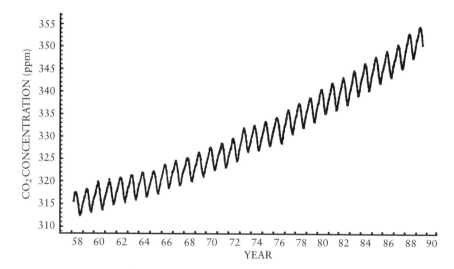

Figure 13.3. Concentration of atmospheric carbon dioxide 1958–1989 at Mauna Loa Observatory, Hawaii.

factors, not affected by human activity, may also play a major role. There is, however, considerable concern about the question of global warming and the potentially harmful effect of excess atmospheric carbon dioxide. The question became such a public concern that in 1998 some 160 countries agreed in the Kyoto Protocol to limit and, in the industrial world, to decrease carbon emissions. To achieve this, however, will be difficult, and it could cause great economic hardship.

How can the goal of decreasing carbon dioxide emissions be achieved with a growing world population and commensurate increasing energy needs? Fossil fuels are still the predominant source for energy. Whereas clean hydroelectric, solar, wind, and other alternate energy sources would be an answer, they can only amount to a modest part of the world's overall energy use in the foreseeable future. Atomic energy is clean in respect to greenhouse gas emissions but has serious problems. Nevertheless, from what we know today, we will have no choice but to use nuclear fission energy in the twenty-first century on a massive scale. We should, however, make it safer and solve the radioactive waste problem (either by improved disposal methods or by new atomic reactor technology). The needed nuclear fuels (uranium, thorium) are themselves not unlimited, but breeder reactors or eventually even controlled fusion may give mankind its needed long-range energy freedom.

The control of carbon dioxide emission from burning fossil fuels in power plants or other industries has been suggested as being possible with different methods, of which sequestration (i.e., collecting CO_2 and injecting it to the depth of the seas) has been much talked about recently. Besides of the obvious cost and technical difficulties, this would only store, not dispose of, CO_2 (although natural processes in the seas eventually can form carbonates, albeit only over very long periods of time).

For my part, although I may be somewhat of a visionary, I see a solution to the problem by chemical recycling of excess carbon dioxide emissions into methyl alcohol and derived hydrocarbon products.

In photosynthesis, nature recycles carbon dioxide and water, using the energy of sunlight, into carbohydrates and thus new plant life. The subsequent formation of fossil fuels from the biomass, however, takes a very long time. We cannot wait for this natural process to take its course and must find our own solution. The question I raised was, can we reverse the process and produce methyl alcohol and derived hydrocarbons by chemically recycling carbon dioxide and water? The answer is yes, but to achieve it in an environmentally friendly and practical way represents a major challenge.

The average carbon dioxide content of the atmosphere, as mentioned, is very low ($\sim0.035\%$) and, therefore, it is difficult to recover CO_2 from air economically (although improved membrane separation techniques may eventually allow this). One must admire nature even more, because plants seem to be able to do this effortlessly. However, CO_2 can be readily recovered from the emissions of power plants burning carbonaceous fuels or from fermentation processes, calcination of limestone, or other industrial sources containing higher concentrations. To convert carbon dioxide subsequently to methyl alcohol and derived hydrocarbons chemically, hydrogen is needed. The water of the seas is an unlimited source of hydrogen, but it must be split (generally by electrolysis), which necessitates much energy. The availability of safer and cleaner atomic energy, as well as alternative energy sources, eventually will provide this. Use of photovoltaic solar energy, for example, is a possibility in suitable locations. Energy of the wind, waves, and tides can potentially also be used. For the present, however, because we still cannot store electricity efficiently, use of the excess capacity of our existing power plants (either burning fossil fuels or using atomic energy) in their off-peak periods represents a convenient source, because they could produce hydrogen that then would be used to catalytically reduce CO_2 to methyl alcohol and derived hydrocarbons. This would allow reversible storage of electricity and at the same time recycle carbon dioxide, not only to mitigate global warming but also to provide an unlimited renewable source for hydrocarbons and their products when our fossil fuel reserves are being depleted.

Besides chemical catalytic reduction of carbon dioxide with hydrogen, which is already possible in the laboratory, we are exploring a new approach to recycling carbon dioxide into methyl alcohol or related oxygenates via aqueous electrocatalytic reduction using what can be called a regenerative fuel cell system. The direct methanol fuel cell

(DMFC) discussed above works on the principle that when methyl alcohol is reacted with oxygen or air over a suitable metal catalyst it forms CO_2 and H_2O, while producing electricity. Similar to the concept of Grove's original fuel cell based on the reversal of the electrolysis of water, it is possible to reverse the methanol fuel cell process and to convert CO_2 and H_2O electrocatalytically back to oxygenates such as formaldehyde, formic acid, methyl formate, and eventually methyl alcohol, depending on the cell potential used. In this way the aqueous electrocatalytic reduction of CO_2 is achieved. Many problems, not the least that of overpotential, must be solved. In aqueous solutions, carbon dioxide is first reduced to the CO_2^- radical, which is further reduced in the presence of water to $HCOO^\bullet + HO^\bullet$ and then to $HCOO^-$. Although the standard reduction potentials for the various CO_2 reduction reactions are small, the large overpotentials observed on metals are most likely due to the formation of the radical anion intermediate. Improvements in electrocatalytic reduction of carbon dioxide continues to lower the overpotential on metals, with the goal of maintaining high current efficiencies and densities. In any case, the regenerative fuel cell process has substantial promise.

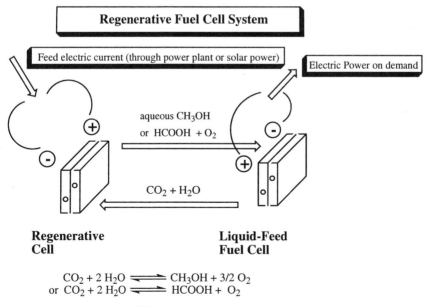

Regenerative Fuel Cell System

Feed electric current (through power plant or solar power)

Electric Power on demand

aqueous CH_3OH
or $HCOOH + O_2$

$CO_2 + H_2O$

$⊕$ $⊖$ $⊕$

**Regenerative
Cell**

**Liquid-Feed
Fuel Cell**

$$CO_2 + 2\,H_2O \rightleftharpoons CH_3OH + 3/2\,O_2$$
$$\text{or } CO_2 + 2\,H_2O \rightleftharpoons HCOOH + O_2$$

Overall Rections

The concept of the reversed fuel cell, as shown schematically, consists of two parts. One is the already discussed direct oxidation fuel cell. The other consists of an electrochemical cell consisting of a membrane electrode assembly where the anode comprises Pt/C (or related) catalysts and the cathode, various metal catalysts on carbon. The membrane used is the new proton-conducting PEM-type membrane we developed, which minimizes crossover.

A regenerative fuel cell system can also be a single electrochemical cell in which both the oxidation of fuels (i.e., production of electric power) and reduction of CO_2 (to obtain fuels) can be carried out by simply reversing the mode of operation.

The conventional electrochemical reduction of carbon dioxide tends to give formic acid as the major product, which can be obtained with a 90% current efficiency using, for example, indium, tin, or mercury cathodes. Being able to convert CO_2 initially to formates or formaldehyde is in itself significant. In our direct oxidation liquid feed fuel cell, varied oxygenates such as formaldehyde, formic acid and methyl formate, dimethoxymethane, trimethoxymethane, trioxane, and dimethyl carbonate are all useful fuels. At the same time, they can also be readily reduced further to methyl alcohol by varied chemical or enzymatic processes.

The chemical recycling of carbon dioxide into usable fuels provides a renewable carbon base to supplement and eventually replace our diminishing natural hydrocarbon resources. Methanol (or dimethyl ether), as discussed, can be readily converted into ethylene or, by further reaction, into propylene.

$$2\,CH_3OH \xrightarrow{-H_2O} CH_3OCH_3 \xrightarrow{cat.} CH_2{=}CH_2 + H_2O$$

$$CH_2{=}CH_2 + CH_3OCH_3 \xrightarrow{cat.} CH_2{=}CHCH_3 + CH_3OH$$

Ethylene (as well as propylene) produced from carbon dioxide subsequently allows ready preparation of the whole array of hydrocarbons, as well as their derivatives and products that have become essential to our everyday life. Whereas the nineteenth century relied mostly on coal for energy as well as derived chemical products, the twentieth century greatly supplemented this with petroleum and natural gas. With our limited fossil fuel resources in the twenty-first century, chemical recycling of carbon dioxide can increasingly play a role and provide a renewable base for our hydrocarbon needs. I emphasize again that the chemical recycling of carbon dioxide of course necessitates much energy, which, however, can be obtained from atomic, solar, or other alternative sources when our fossil fuel sources diminish.

The world's energy and material sources have difficulty in keeping up with our still rapidly growing population and increasingly technological society. New and more efficient ways are needed to satisfy demands so that a reasonable standard of living can be provided for all. Society's demands must be satisfied while safeguarding the environment and allowing future generations to continue to enjoy planet Earth as a hospitable home. To establish an equilibrium between providing for mankind's needs and safeguarding and improving the environment is one of the major challenges of mankind. As a chemist, I find it rewarding to have been and to continue to be able to work toward this goal.

· 14 ·

Gone My Way

Coming to the end of my recollections of a full and rewarding career, I look back with satisfaction, but even more with gratitude. I had the good fortune that my career spanned the exciting time for science that was the second half of the twentieth century, in which I was able to actively participate. I would like to reflect on some of the beliefs and principles that always guided me. I cannot claim to have had great insight or to have wisely planned my career. My ways developed over the years mostly from my own experiences and circumstances, coming quite naturally. I am only too aware of my limitations and shortcomings. Thus I do not want to give the impression that I consider "my ways" always to have been the right ones or even the most reasonable and certainly not the "safe" ones. We all are born with an inherited nature, but we are also affected by our environment, by our experiences, fortunes, and misfortunes, and, most of all by our free will and ability to overcome adversities and to succeed.

The education many Hungarian-born scientists received in our native country provided the foundation on which our subsequent scientific careers were built. At the same time, it must be remembered that most of us who became somewhat successful and noted did so only after we left Hungary. Denis Gabor (Nobel Prize in physics 1971 for holography), John Neumann (pioneer of computers and noted mathematician), John Kemeny (mathematician and developer of early computer programs), Leo Szilard (physicist), Edward Teller (atom physicist), Eugene Wigner (Nobel Prize in physics 1963 for theory of atomic physics), George Hevesy (Nobel Prize in chemistry 1943 for isotopic tracers), like myself, all received recognition while working in the West.

Only Albert Szent Gyorgyi received his Nobel Prize (physiology or medicine 1937 for biological combustion processes with reference to vitamin C) while working in Hungary, but he also spent the last four decades of his career in the United States. Perhaps a small country could not offer its scientists opportunities for growth and work at home. Notably, there is no woman scientist on the list, perhaps telling something about the gender-separated educational system of the past and the bias against giving women an equal chance for a scientific career. I remember that in my university days there was not a single girl in our class! Hitler and Stalin also greatly contributed to the flight from Hungary. At the same time, I know that there were many highly talented contemporaries of ours who never left Hungary but somehow did not develop to their real potential. In any case, an isolated, small, and rather poor country, despite all its handicaps, turned out a remarkable number of excellent scientists, mathematicians, engineers, composers, musicians, filmmakers, writers, economists, and business leaders.

I hope that working conditions and opportunities in Hungary will change and improve. In 1989, Hungary rejected four decades of Communism (even if the last decade of "goulash communism" was only a milder left-over version) and started its development as a free, democratic country. It is not easy to achieve fundamental changes in a short time, but remarkable progress has been made. In 1984, when I first returned to Hungary after 27 years, the country was just beginning the changes that greatly accelerated after the turn to democracy in 1989. Hungarian science, which was treated relatively well by the Communist regime, was based substantially on a large framework of research institutes (run by the Academy of Sciences as well as state ministries and industries). At the same time universities were weakened, because much of the support for research was directed away from them toward these institutes. The process of reestablishing and strengthening university education and research has begun. The research institutes are reorganized into a more manageable scope and size, with combining some and even privatization of others, at the same time establishing a close cooperation with universities and private industries. Of course, in the midst of fundamental economic changes and inevitable hardships, this is not an easy task. Many younger people, this time mainly for economic reasons and better opportunities, are looking to the West. Whereas some experience in a foreign country is beneficial for the development of any scientist, it is hoped that Hungary increasingly will be able to offer opportunities for its young scientists that will induce them to stay or return home.

It is only natural that Hungarian-born scientists around the world who owe much to their heritage and the educational system of their native country are trying to help to the best of their ability to bring about these changes and improvements. For my part, I have in the past decade welcomed a series of younger Hungarian colleagues for stays of up to two years in our Institute (primarily from the University of Szeged, from which fine colleagues such as Arpad Molnar, Imre Bucsi, Bela Török, and Istvan Palinko came) and we are maintaining continued contact and research cooperation with them after their return home. I also helped to endow a research award of the Hungarian Academy of Sciences (which generously named it after me) given in recognition of the achievement of younger chemists. Judy and I visit Hungary perhaps every other year for some lectures and events (the Nobel Prize has a great attraction in a small country). I always try to emphasize to young people the fundamental need to get a good education and have faith in the future.

To the frequently asked question of what could I have achieved if I had not left in 1956, I have no definite answer. I believe, however, that we should look to the future and not second-guess the past. It is the future that really counts, and the key to the future (not only in Hungary but worldwide) is a good education. Only with a good education can the younger generation succeed, but of course circumstances, desire, and ability (as well as luck) are also important.

Once I had decided on a career in chemistry, I was determined rather single-mindedly to make a success of it. I sometimes think about what would have happened had I chosen a different occupation or field. Having a rather competitive nature, I could probably have done reasonably well in a number of other areas. Certainly for some fields you must be born with a special talent. Musical talent, artistic ability, business acumen, leadership ability, and vision can be further developed, but there must be a given inner core. In creative science (and I empha-size "creative," not necessarily revolutionary in Thomas Kuhn's sense) there also must be some of this, but probably the acquired aspects are equally or more important. The genius of a Galileo, Newton, or Ein-stein is not questioned, but there are very few trailblazers who open up entirely new vistas. In mathematics, not unlike in music and the arts, you must have a talent born with you, and this is also the case in fields where mathematical thinking plays a major role. In contrast, in my field of chemistry, I believe it is not so much a specific inborn talent that matters but the ability for unconventional (perhaps it can be called creative) thinking and the ability to realize the significance when you come across something new. You also must have the ability and the staying power to subsequently explore it.

I strongly believe in the central role "multifaceted chemistry" plays in bridging other sciences and the ways it affects all aspects of our life. In a guest editorial for the journal *Science* in December 1995 I wrote:

> Humanity's drive to uncover the secrets of life processes and to use this knowledge to improve human existence has led to spectacular advances in the biological and health sciences. Chemistry richly contributes to these advances by helping to increase our understanding of processes at the molecular level, and it provides many of the methods and techniques of biotechnology. However, chemistry is not just an adjunct of biology and biotechnology. It is and always will be a central science in its own right.

Chemists make compounds and strive to understand their reactions. Chemical synthesis, coupled with biotechnology, is well on its way to being able to reproduce many of nature's wonderful complex com-pounds and also to make unnatural ones. Although I started out as a natural product chemist coming from the Emil Fischer school, my own interests led me eventually to the chemistry of hydrocarbons. Hydro-carbons are essential to our everyday life because they make up petro-leum oil and natural gas. Hydrocarbon fuels generate energy and elec-tricity, heat our houses, and propel our cars and airplanes. They are also the raw materials for most man-made substances, ranging from plastics to pharmaceuticals. What nature has given us over the eons we are, however, using up rapidly. As our nonrenewable fossil fuel reserves diminish in the twenty-first century, it will be up to chemists to synthesize hydrocarbons on an increasingly large scale and in new and economical ways.

My longstanding research in hydrocarbon chemistry, particularly on acid-catalyzed reactions and their carbocationic intermediates, has also yielded practical results. We were able to improve widely used acid-catalyzed industrial processes and to develop new ones. For example, in the production of high-octane and oxygenated gasoline, toxic and dangerous acids such as hydrofluoric acid are used in refineries. We found it possible to modify these by complexation with additives to make them substantially less volatile and thus much safer. New generations of highly acidic solid catalysts were developed for efficient and safe processes. So were environmentally adaptable and safe gasoline and diesel fuel additives, which allow cleaner burning and less pollution. To find ways to produce hydrocarbons by synthesis, as discussed in Chapter 13, we are working to convert natural gas directly to liquid hydrocarbons and their oxygenated derivatives. Furthermore, I believe it will be possible to convert excess carbon dioxide from the atmosphere into hydrocarbons if we can find sources of energy (safer atomic and other alternate sources, including solar energy) needed to produce the necessary hydrogen. We will then be able to supplement nature's photosynthetic recycling of carbon dioxide and provide for our hydrocarbon needs after we have exhausted our fossil fuels. The relevant chemistry is being developed in laboratories such as ours. Significantly, recycling carbon dioxide will also mitigate global warming caused by excessive burning of fossil fuels or other hydrocarbon sources.

Chemistry, I fully realize, does not always enjoy the best of reputations. Many chemical plants and refineries are potentially dangerous and pollute their surroundings. At the same time, however, their products are needed, and our society enjoys a high standard of living that few are willing to give up. Our advanced way of life is in no small measure helped by the results of chemistry. Chemistry should and I believe will be able to bring into equilibrium providing for our needs and responding to societal environmental concerns.

I have tried in this book to reflect on the search for the personal "magic chemistry" that led me through my career. Chemistry attracted me through its wide scope and many possibilities for exploring fascinating new areas. I enjoyed (and still do) the challenge and excitement of entering new areas from time to time and always found in them sufficient terra incognita for exploration of the unknown and unconventional.

The Indian-born physicist Subramanyan Chandrasekhar (Nobel Prize 1983) had a personal style of research, which I learned about only recently, that seems to parallel mine. He intensely studied a selected subject for years. At the end of this period he generally summarized his work and thoughts in a book or a comprehensive review and then moved on to something else. He also refuted Huxley's claim that "scientists over 60 do more harm than good" by sharing the response of Rayleigh (who was 67 at the time) that this may be the case if they only undertake to criticize the work of younger men (women were not yet mentioned) but not when they stick to the things they are competent in. He manifested this belief in his seminal work on black holes, on which he published a fundamental book when he was 72, and his detailed analysis of Newton's famous *Principia* published when he was 84, shortly before his death. I have also written books and reviews whenever I felt that I had sufficiently explored a field in my research and it was time to move on; this indeed closely resembles Chandrasekhar's approach (vide infra).

Much is said about at what age scientists become unproductive and should quit. As they grow older many cease doing research. Others continue but try to overreach in an attempt to produce something "lasting" and to tackle problems, often with embarrassing results, outside their field of expertise. It is said that physicists do their best work in their twenties or thirties, whereas in other fields age is less of a detriment. I myself never felt influenced by such questions or considerations. Maybe I am lucky, but writing this at age 73 I still feel the drive and satisfaction of pursuing my search to understand and explore new chemistry. If this goes away, I fully intend to retire. There are many other satisfying things one can pursue and enjoy besides active research.

It has been my way to pursue my interests free of any a priori judgment of whether the topic was popular or well received at the time. Such considerations, at any rate, are temporary and tend to change with time. It is also more fun to be one of the leaders in a new area, regardless of its popularity, than just to follow the crowd or be part of a stampede in a "boom area" (which frequently tends to bust eventually). Of course, the price you pay when you go your own way is that not only that recognition of your efforts may be long in coming (if it comes at all) but also it is more difficult to get support for your work. An inevitable price for the freedom of academic research is the necessity to raise support for it (in chemistry, mostly to support your graduate students and postdoctoral fellows). In our American system, government agencies such as the National Institutes of Health (NIH), National Science Foundation (NSF), Department of Energy (DOE), and various Department of Defense (DoD)-related research offices [i.e., Army, Navy, Air Force, DARPA (Defense Advanced Research Project Agency)] are some of the major sources of support. These research agencies are basically independent of each other and are even in some friendly competition. In my view, this helps to keep American science vibrant, highly competitive, and active. In most other countries there is generally a single central research support agency. If somebody in that agency does not like your type of research, that is generally the end of it. In our system, although the competition for research support is fierce, the fact that there are a number of agencies that could be interested in supporting your work offers more flexibility. This does not mean, however, that you do not need to put a major effort into your research proposals. You also must have flexibility on occasion to adjust some directions of your planned work to match it better with the interest of a specific agency. Proposals, as a rule, are strictly peer reviewed. One could disagree on whether this system is the best, but I am a firm believer that, overall, it does work out and nobody has come up with a better system yet. Of course, there are people who have a talent for writing proposals and show great "salesmanship." Others are less adept at it, although not necessarily lesser scientists.

It is particularly difficult for young people to break into the system and for older ones to stay in it. Years ago I suggested an approach to help the former. If someone gets a junior faculty appointment at a reputable academic institution, this should give assurance that he/she is sufficiently qualified to justify the promise (based on both background and also promise for independent research). Therefore, that young researcher should be trusted with an initial grant giving substantial freedom to pursue the chosen research. Most established researchers when they write proposals in fact know part of the outcome based on work they have already done in the context of their previous funding. They are thus in a much stronger position to suggest future work with a high probability of success. In a way, under my suggested approach we would give our younger colleagues an initial credit. At the end of the grant period, the work done would be reviewed as to whether it produced tangible results. If so, it would warrant extension of support (or denial if the results were insufficient). Needless to say, my idea has not been considered, but I still believe it has merit.

Concerning the other end of the scale, it is difficult for someone of my age to be entirely impartial. Aging, however, is not necessarily something to be measured entirely by calendar years. Some stay very productive and active for a long time, whereas others, for whatever reason, tire of research earlier but frequently coast along for many years on safe, but not necessarily exciting, extensions of their previous work. I have no easy formula to offer, but I believe that age per se should not be used to discriminate. There is, of course, the valid point that having only limited funds and research facilities it is necessary to open up possibilities for younger researchers, even if this means closing off some for the older ones. I do not want to question the validity of such a consideration, but, of course, nature provides its inevitable renewal cycle. What I believe is that ways could be found to act more tactfully, with some human consideration of elderly colleagues who still can make valuable contributions and compete well on merit with their younger colleagues.

To recall a personal experience, my research on carbocations and their related chemistry, because of its significance and relevance to biological systems and processes, was supported in part by NIH for nearly 30 years, for which I am most grateful. In 1994, however, my support was not renewed. I was told that it was judged that no further significant or relevant new findings could be expected from my work. This happened at the time when I was making probably my most interesting and significant research discoveries on the new concept of superelectrophilic activation (Chapter 12). As it turned out, this has substantial general significance to biological systems and even enzymatic processes, among others. Coincidentally, shortly after I was turned down by NIH I won the Nobel Prize and received a congratulatory letter from the NIH Director saying how proud they are of supporting my work (in the present tense). Bureaucratic mix-ups of course happen, but at the time I was not amused. In the long run, however, things worked out well. The nonrenewal of my support reinvigorated my research effort. It is always rewarding, instead of brooding over some adversity, to find a way to overcome it. I was able to continue my work with continued support from other agencies as well as private supporters and foundations, who still had confidence in my work. I was able to prove that there was still much new pioneering chemistry left for me to discover and that the judgment of the futility of my proposed research was premature. As a matter of fact, I believe that my post-Nobel research has turned out to be quite exciting and productive.

Whereas we have eliminated most discrimination in our society, age discrimination is a sensitive topic not frequently discussed. Officially, of course, it is not admitted, but de facto it very much exists. I have known many outstanding scientists, including Nobelists, who were forced to give up research at a time when they still felt that they could make significant contributions. Finding some ways to support them (which on the overall scale of funding would be not significant) is something worthwhile to consider. Some of our younger colleagues sitting in sometimes harsh judgment on review panels may not always realize that in years to come they will find themselves on the other side of the question, and I am sure by then their views may be different.

Besides support by governmental agencies there is also support of research by industry, varied foundations and charitable organizations, and individual donors. This all adds to the diversity of our research enterprise and its vitality. The U.S. National Academy of Sciences itself does not support research to any significant degree. As a private organization it has no governmental budget or extensive private means. It, however, undertakes studies and reviews to assist the government and increasingly tackles significant, sometimes controversial, questions of general interest on its own. This is different from some other countries, in which state-supported academies have their own research institutes. This was (and still is) to a large degree the case in the former Eastern Bloc countries. In a different context, governmental support allows organizations such as the German Max Planck Society to maintain their numerous institutes for basic research or the Fraunhofer Society to maintain their more industrially oriented institutes. The closest we come in the United States to these institutes are the NIH Institutes and the National Laboratories, the latter mainly concerned with energy and military research.

With my European background, I was when I came to America and still am impressed by the rather loosely organized, more decentralized way of research support. Of course, even in a great country like ours resources are not limitless and inevitably prevailing trends of research set priorities. In my field of interest the 1970s and 1980s were a period when, after two oil crises, research on hydrocarbon fuels and their synthetic preparation had significant public interest and support. Catalytic research in its many aspects was heavily pursued and considered a national priority.

This changed significantly in the 1990s. The unprecedented prosperity and economic upturn induced the public to believe that concerns over our limited oil and gas reserves and our energy needs in general are of no real significance. The national science priorities heavily turned to the biological-life-health fields with strong emphasis on biotechnological developments. The development of electronics and computer-based technologies and industries, including development of new materials, were also spectacular and are continuing, but they are significantly funded by private industry and enterprise. I certainly do not want to minimize the enormous achievements and significance of these fields. It is obvious to me, however, that to maintain and continue our technology-driven development for the benefit of all mankind, we cannot neglect such basic questions as how we will provide the essentials for mankind's everyday life, including all the energy and materials we need. We also must maintain or restore a clean and safe environment for all the inhabitants of Earth, who may reach 10 billion by the middle of the twenty-first century. To this end, it is essential to put more emphasis on research in areas such as those connected with energy, non-renewable basic resources such as hydrocarbons, as well as solutions to environmental problems (as contrasted with only trying to regulate them).

I have discussed my efforts in some of these areas in starting to build a small university-based research institute on my move to Los Angeles in the late 1970s. I am proud that we have not only achieved this, but kept it going for nearly a quarter of a century. It is a privately supported institute, with no direct governmental (federal or state) funding. It is still a wonderful aspect of America that if you have a vision and work hard for it you can achieve your goals. The Loker Institute raised all the funds for its existence from private donors, friends, and benefactors, and established some endowments to support pioneering work in new, high-risk areas and to provide the framework in which our faculty can compete successfully for traditional, competitive research support from agencies for their specific research projects. The Institute's work itself contributes to some degree to its operation through income from patents and intellectual property obtained from our discoveries and other efforts. The Institute has associated with it a number of endowed chairs, giving our faculty further support and well-deserved recognition. We also have been able to endow our symposia and an annual lecture in the field of our broad research interests, as well as our own essential small library. Friends, first of all Katherine Loker, as well as the late Harold Moulton, and other supporters and foundations gave generously. Carl Franklin continues to be particularly supportive in raising funds for the Institute, while himself also being one of our supporters. I learned much from him, for example, that an essential aspect when you try to raise support is to contribute yourself to your best ability. It could be said more bluntly as "put some of your money where your mouth is." Nothing impresses others to contribute to your cause as much as your own example. Alfred Nobel may be even pleased that part of his prize money found its way to support our Institute. Judy and I are extremely grateful to our adopted country for the opportunities it gave us and also to the University of Southern California, which became an integral part of our life. Returning some of our good fortune thus is a privilege and a small expression of our gratitude.

I was fortunate that during my career I always had varied, broad interests and was able to explore a diverse palette of research topics cutting across conventional lines of chemistry. To follow my own interests and ideas came naturally. I always liked marching to my own drummer and was never scared to divert from the safe and well-traveled paths of "normal science" (in Thomas Kuhn's sense) to terra incognita (which many feel is unwise and they keep away from it). Sometimes I indeed disregarded what were the recognized limits of the terra firma of established knowledge. I reached outside them to explore unknown areas and ideas of my own, despite the cautioning and even disapproval of some of my respected, more experienced and knowledgeable peers and colleagues. I had the good sense, however, to keep my eyes open and was prepared for the unexpected and unconventional. When these came my way, I was able to recognize and follow them up. I also asked myself, when I found something interesting and new, whether it could be used for something useful. Because I spent some years in industry and maintained an interest in the practical, it was never difficult for me to freely cross the line between the basic and more applied aspects of my research.

When considering a new research project, I always found it very worthwhile to pause and question, assuming everything I was expecting to achieve would come through (and, of course, all researchers know that in reality this practically never happens), what the significance of the result would be. I remember hearing Frank Westheimer, an outstanding Harvard chemist, recall his encounter as a young graduate student with James Conant, the eminent chemistry professor under whose guidance he started his work but who subsequently, upon becoming Harvard's president, moved away from chemistry. He inquired how his former student was doing in his research. Westheimer proudly told him what he had done and what he expected to do further. Conant politely listened and then said, "Frank, have you considered that even if you achieve all this, at best it may be just a footnote to a footnote in chemistry." He learned to choose his projects carefully, and so did I.

It is very useful not only to consider how to achieve one's research project, but also to ask early on the question of its worthiness and significance if it were successfully carried out. It is the nature of research that once you are fully engaged in a project it becomes a very personal involvement. Being too close to it tends to influence your judgment, and the day-by-day effort to push ahead despite inevitable difficulties and disappointments becomes dominant. The question of whether the project, even if successful, would be worth the effort to continue at this stage is usually not considered. I believe this frequently leads to the proliferation of otherwise solid but not necessarily really significant or original research efforts. Even if you are blessed with a somewhat creative mind producing new research ideas, the most difficult part is to sort them out and decide which of them is really worth being followed up. This is even more of a responsibility when you are an advisor of your students or a research director in industry. The enthusiasm, hard work, and fresh ideas of your graduate students and postdoctoral fellows are of course the backbone of any research effort in your laboratory. It is, however, your responsibility to set the overall direction and to keep the work focused. In the course of any research project, unexpected and new observations can come along. You must keep your eyes open to recognize their significance, which can lead to new and frequently more significant directions than originally foreseen. The most difficult decision, however, is to know when to stop pursuing ideas that turn out to be unrealistic or unproductive. You must then redirect the work, learning from your disappointments and mistakes (some of which in retrospect you perhaps should have foreseen).

In some ways, this is also the case in industrial research, which I have also experienced. Projects frequently could be stretched out for a long time, even producing solid results, without, however, leading to real breakthroughs or solutions. At the same time, however, it is also essential to know how to stay the course and stick with your project and not give up prematurely because of some unavoidable disappointments, faulty starts, or and other difficulties, without giving it your best shot as long as it is reasonable. I do not have a magic recipe for how to direct successful research. My experience, however, tells me that you should be able to find the right balance to make your decisions. It is also necessary to have some faith to follow within reason your intuition and imagination, for which you eventually can be rewarded if some of your ideas indeed turn into successful reality.

As I mentioned, it is necessary to have a realistic degree of belief in your ideas and the ability to follow them through. If you yourself do not believe in your project, how can you convince your students and colleagues to give it their best effort? Building up their confidence is also essential. I always started my graduate students on some smaller project with a higher degree of probability for success. Once they gained some confidence in their research they were ready to move on to higher-risk, more challenging problems, without easily being discouraged by disappointments.

In my career I have experienced many instances when my self-confidence was seriously tested. This is inevitable if you are not just pursuing regular, safe science. Because from my earliest days in research I preferred to go my own way searching the unexplored and unconventional, disappointments were unavoidable. When later on I inadvertently got involved in some controversies (most notably the nonclassical ion controversy), the general attitude of my colleagues was not to take any chances and to "sit out" the controversy until overwhelming evidence led to its resolution. However, this was not my way. Once I was convinced of the validity of our relevant research results, I was prepared to stand up for my views. The personal attacks and criticism this frequently brought about were not easy to take at times. If you believe that your work is correct (and, of course, you must make sure through repeated, unequivocal testing that your experimental results are correct), then you should be able to stand by your opinion, however lonely or unpopular it may be at the time.

Over the years, as mentioned, I followed the practice that, whenever in a specific area of my research I felt that I had substantially achieved my goals and that it was time to consider shifting my emphasis elsewhere, I wrote (or edited) a book or comprehensive review of the field. My books are listed in the Appendix for interested readers who want to obtain more information or details, as well as relevant literature references.

My initial excursion into writing was in Hungary, where my university lectures on theoretical organic chemistry (really physical organic chemistry) were published in two volumes in 1955. A substantially revised German version of the first volume, written in 1956, was published in Berlin in 1960 (the second volume, however, was never realized due to circumstances). With few exceptions (as indicated) all my subsequent books were published by my publisher Wiley-Interscience in New York, with whom I still have a most rewarding, close relationship. It started in the early 1960s when I met Eric Proskauer, who at the time was running Interscience Publisher, and Ed Immergut, who was his editor. Later, Interscience was bought out by John Wiley and we continued our relationship, with Ted Hoffman becoming my longtime editor and friend with whom it was always a pleasure to work. Barbara Goldman and now Darla Henderson took over to continue a rewarding relationship. I also consulted for Wiley for many years concerning their organic chemistry publishing program and was able from time to time to suggest new authors and projects. I remember, for example, the start of Saul Patai's project, *The Chemistry of Functional Groups*, which later, with the involvement of Zvi Rappoport, developed into a truly monumental series. To keep the volumes from becoming obsolete, at one point I suggested adding supplemental volumes to upgrade the discussed topics. These turned out to be very useful.

The Fiesers' *Organic Reagents* series, I remember, was originally Mel Newman's idea, but he himself was not interested in carrying it out. The Fiesers made it a real success. After Louis' death Mary carried it on for many years, and now, at my suggestion, Tse Lok Ho, a friend and former associate, continues the series. Speaking of the Fiesers, they used to have a photograph of their cats adorning the front of their books. Being a dog lover, I felt that canines deserved equal treatment and for nearly three decades the pictures of our Cocker Spaniels, Jimmy and Mookie, brightened for the benefit of other dog lovers an otherwise wasted front page of my books. As my love and pride became my grandchildren, it is only befitting that their picture adorns the front of this book.

Writing or editing scientific books is always a rewarding experience, with the authors gaining most from the systematic study and review

Two faithful friends. Jimmy 1969–1984 and Mookie 1985–1998

of their field. Even more important is to be able to evaluate and sort out what is really important to be discussed and how to present it. It is also frequently a stimulating process, because during review of a field, one frequently realizes important yet still unresolved problems or gaps worth studying or even gets ideas for new aspects to be explored. My books were always on topics to which I felt I made significant contributions. I was thus able to incorporate my own work and views and place them in proper perspective. I also hope that if future readers look up my books in libraries (assuming that printed books will survive and not be replaced entirely by electronic systems) they will find them still of some interest and use.

As rewarding as scientific publishing and writing is from an intellectual point of view, its monetary rewards are generally minimal (except for authors of successful college textbooks and advanced texts which are also used as such). I mention this without any regret or complaint. Royalties or honoraria never crossed my mind as a factor in any of my writings. Certainly a similar effort put into any other endeavor would have been much more rewarding financially. There is, however, the satisfaction of writing something you believe will be of some value to your field and also the sheer joy of seeing your book finally published, something that only those who have experienced it can fully appreciate. It is in a way like seeing your newborn child for the first time. Electronic publishing, however promising, I feel will not be able to give the same inspiration and incentive to authors to go through the long, demanding task of months or years to produce a comprehensive scientific review or book (or, for that matter, even research papers). Of course, I may be wrong and only a holdback of my generation, but I myself will not be around to find that out.

Besides my research I always considered teaching an equally significant part of my professional life. As a parent and grandparent, too, I believe that a good education is the single most significant asset we can give to our children. In a commencement address I gave at the University of Southern California in May 1995 I reflected on this.

... There is nothing more important and necessary in facing a dynamic new century than to be able to compete effectively in the arena of real life and to be able to offer the knowledge and skills demanded to succeed. We have undergone amazingly fast technological development in the relatively short period of two centuries since the industrial revolution. There is no single aspect of our life which has not been touched and fundamentally changed by it. Progress is not only continuing but is accelerating. Just recall some of what happened in our own 20th century: the general use of electricity, dawn of the atomic age, fundamental changes in transportation (think about cars, planes, etc.) in communications (telephone, radio, television, satellite systems, FAX), the multitude of emerging new miracles of electronics, the enormous impact of the computer in all aspects of our life. We take all of them for granted as we near the end of the century. Whereas Science laid the foundations through developing fundamental knowledge, it was application by technology (engineering, manufacturing) which led to practical uses resulting in our highly technological oriented society. Of course, human knowledge and endeavor are broader in scope than just science or technology, and the arts and humanities much enrich our life. In life the most significant foundation for all of us is education and training which allow to be prepared for a productive and rewarding life. Nobody will be able to compete effectively at any level of the work force without the education and skills needed in the 21st century, which is just around the corner. Technological advances will inevitably result in further increases in productivity, freeing more time for the individual to be able to spend in a meaningful ways.

. . . Education is a pyramidal edifice—colleges and universities must build on the foundation laid by our primary and secondary schools. On their own, as fine as our institutions of higher education may be, they can only fail and be bogged down by flaws and shortcomings in earlier education. Schools further can not do the job alone. It's the family which provides the surrounding and help that together with the schools give young people in their crucial years of growing up and maturing the environment essential to obtaining a solid education. One without the other cannot succeed. Society of course should and indeed try to help whenever the need arises, but neither Government nor private efforts will ever be able to replace the role and significance of the family. . . .

I have always enjoyed teaching and direct contact with my students. Teaching chemistry varies in its approach at different levels. Early courses for nonmajor undergraduates should provide sufficient, but not in-depth, introduction emphasizing the wonderful, magic world of chemistry and its broad applications as well as its significance to the other sciences. Regrettably, chemistry frequently is still thought of as a rather dry discipline based primarily on physical laws. However, chemistry can be taught, even while acknowledging physical principles, as a vibrant, exciting topic with much relevance to and examples from our everyday life and the challenges and problems of mankind. Linus Pauling taught such a course at Caltech for years, and there is indeed a shift these days toward teaching more relevant, interesting chemistry courses.

With the advance of recorded or "on-line" courses readily transmitted through the web or other electronic media, some argue that the need for direct, live teaching diminishes. This may indeed be the case for service courses or specific engineering or other professional updating or training. I believe, however, that nothing can replace direct personal contact between students and teacher, which, in addition to information, also can give motivation and even inspiration. Furthermore, a good teacher should give much more than simply a recitation of a book or lecture notes or a prerecorded lecture. If these were sufficient, there indeed would be no need for the direct classroom experience. However, attending lectures in which the topic is presented in a personal, dynamic, and, frequently, new way is a lasting experience. Furthermore, having direct contact with professors who, through their own knowledge and experience, can make the topic come alive for their audience is also a unique experience. Pauling's freshman lectures and Feyman's physics lectures at Caltech come to my mind, as well as Woodward's Harvard lectures, in which he, in his precise manner, artistically built up on the blackboard edifices of organic synthesis. These are just a few examples of many remarkable science courses. If you actively participate in and contribute to science and are a dedicated teacher who continues to study and keep up with the field, teaching and lecturing to your students and other audiences will give them much more than dry recitals. This is and should be the experience that students will remember long after they pass through your classroom.

At more advanced levels, chemistry is learned by doing it. A young surgeon does not learn everything in the classroom or through anatomy dissections and, these days, electronic simulation or watching surgery being performed. Eventually he/she must step up to the operating table and, under the guidance and supervision of practiced peers, begin to really learn surgery by performing it. The same is true for chemical research, which can be learned only by doing it. Laboratory education and practice prepares one, but nothing comes close to the real experience. I have never felt, therefore, that teaching and research can be considered separate entities. Not only are they inseparably mashed together, but the learning process never really ends. A good research laboratory is not only a place to carry out research but also a challenging and inspiring environment that fosters continuous intellectual interaction and self-education. We never cease to learn, not only by keeping up with the scientific literature but also through discussions and interchange with our colleagues, lab mates, and others. Graduate students and postdoctoral fellows, as well as their faculty advisors, learn continuously from each other in an inspiring research environment, much more than any organized course or seminar can provide.

It is not necessarily the faculty itself (however fine it may be) or the excellence of the students (who frequently more than anything make their professors stand out) that makes some chemistry departments really outstanding. It is the conducive, collegial, and at the same time competitive atmosphere that allows scholarly work and research to prosper. It is possible to build new laboratories or institutes, hire good faculty, and provide the prerequisites for their work, but eventual success will depend whether an atmosphere can be developed that brings about the challenging and conducive spirit essential for innovative research and learning. Only this can help turn young, eager researchers into independent scholars capable of pursuing their own work.

An integral part during all my years of research and teaching has been a regular weekly meeting with my group. We discuss our research progress, on which individuals are asked rather randomly to report in an informal way. Questions are raised, and problems are debated. We also discuss some interesting new chemistry reported in the recent literature and occasionally have some more formal presentations (such as reviewing thesis defenses and presentation from the group at upcoming symposia, congresses, etc.). More recently the joint weekly meetings at the Loker Institute of the Olah and Prakash groups have served to mutually benefit both groups. They are very useful regular weekly events that also prepare our younger colleagues well for public presentations, not to mention job interviews.

I am fortunate in that early in my career I got used to lecturing and public speaking without written texts or even notes. For my courses I prepared detailed outlines and selected the material to be discussed, but this was only the framework, which I always changed and updated. This also assures me of never getting bored while giving talks or lectures even on the same topic on different occasions, because I inevitably improvise and never really give the same presentation twice. This brings in new aspects and leaves out others. Because my interests always were (and still are) rather broad, I can lecture on a variety of topics, selecting those most appropriate for the specific occasion. I find it very useful to keep good contact with the audience, preferentially selecting someone for eye contact and, in a way, speaking directly to this person. In my classes, I also found it useful to involve the students through participation in an informal give-and-take fashion. Having a somewhat deep voice that comes through rather well (of course with my unmistakable Hungarian accent) helps to keep my classes awake and to attract my audience's attention. If on occasion some still doze off (a discouraging event for any speaker or teacher), I usually stop and with some humor awaken them, promising to be more interesting and entertaining (which I always try to do). At the end of a lecture I know myself best how I have fared, and the experience helps me to improve. Some of my colleagues spend much more time and concern preparing for their lectures and some even agonize over them. I am, I guess, fortunate that I rather enjoy lecturing and that it does not represent any particular stress for me. Of course, you still must prepare yourself and know your topic well, even more so than when you are not using a prepared text. Answering questions and participating in discussions is particularly stimulating because you can gauge how your presentation and message really got through. You also can learn and get new ideas and stimulation from the comments and questions of your audience.

In chemistry, I think only in English (which is not always the case in other areas). When abroad, I lecture readily and spontaneously in German (although it takes a few days to get back into the swing of it). Interestingly, it is somewhat difficult for me to lecture about chemistry in my native Hungarian (which I speak at home with Judy). Although I was 29 years old when I left Hungary and my mother tongue comes to me naturally, chemistry, as all the sciences, has its own dynamic language. After being away for nearly 30 years, when I first got back to Hungary and opted to give a lecture in Hungarian I found out how much the technical language had evolved (although since that time I manage better). I once spoke French rather fluently, but without practice it deteriorated to the point where I still understand and manage somehow but would not dare to lecture in French (you can make mistakes in other languages, but in French it would be an offense).

I have not kept records of the lectures I gave during my career at seminars, symposia, meetings, and congresses or at different universities around the world. There were many, and although these days I turn down most invitations, I still lecture regularly (but more selectively). Over the years I visited and lectured widely across Europe, and we still take an annual trip to Europe. When our sons were growing up and still willing (increasingly reluctantly), we took them along to show them Europe and expose them to its culture. Since 1984 our visits on occasion also included my native Hungary, generally arranged by a long-time friend Csaba Szantay. In the past decade since Hungary regained its freedom, our visits have become more regular on a biannual basis. In 1995 we met with the President and Prime Minister of Hungary and I was later honored with a Hungarian State Award. I use my travels to renew contact with many friends and colleagues and also to visit with former members of the Olah group scattered around the world.

On a few occasions I have visited Japan (my former postdoctoral fellows there number some 25, and they have an informal association

In the mid-1960s with my "reluctant" sons on a European trip

Hungary 1995 with President Göncz

and meet regularly) and Israel, and once we visited India, where we greatly enjoyed the hospitality of the family of Surya Prakash. We have also visited Hong Kong and Taiwan but never went to mainland China or many other places, including South America and Australia, among others. Judy does not like long flights, and we are also cautious about health problems while in foreign countries. Thus we probably will not

With Prime Minister Gyula Horn

see many more parts of the world, but we still enjoy our travels and visits to familiar places. We are basically not "great travelers," and prefer to stay at home.

To pursue science and keep abreast of my group's work still takes up most of my time and energy. I particularly enjoyed, however, over the years some visiting professorships or lectureships at different universities, which allowed close contact with faculty friends and students alike. I spent a semester at Ohio State in Columbus in the fall of 1963 as a guest of Mel Newman, who offered the use of his laboratory to Judy and me, and we did some rewarding work. I spent the spring of 1965 in Heidelberg as Heinz Staab and George Wittig's guest. In 1969 we spent a most pleasant month at the University of Colorado in Boulder hosted by Stan Cristol. In 1972 I was at the ETH in Zurich hosted by Heini Zollinger in his chemical technology department. In 1973 we were in Munich and had a particularly pleasant time with our friends Rolf and Trudl Huisgen, whom we visit regularly even now. Strasbourg in France was always one of our favorite cities, and in 1974 we spent a month at the university. Jean Sommer, as well as other faculty friends, always made this and other visits to Strasbourg wonderful events (including the culinary wonders of the region). We also visited and stayed on different occasions at the universities of Paris (Dubois), Montpellier (Commeyras, Corriu), Bordeaux, and Poitier (Jacquesy), among others, extending our experience in France. The University of London and particularly King's College represented, besides Durham (where I had close professional and personal ties with Dick Chambers and Ken Wade over the years), a special contact over the years. Since the late 1950s we have visited London annually (where Judy's aunt Alice, with whom she was always very close, lived) and I regularly lectured at King's. The late Victor Gold, as well as Colin Reese and other friends, always extended great hospitality and collegiality, and as a honorary lecturer I always enjoyed my visits and talks. We also enjoyed stays in Italy, in Rome (with the late Gabriello Illuminati and with Fulvio Cacace), in Padova (with Giorgio Modena), and particularly in Milan with my late friend Massimo Simonetta and his wife Mirella. I had close interests and scientific cooperation with Massimo but first of all enjoyed the friendship of the Simonettas. Massimo also introduced me to the Italian chemical industry, and I consulted for some time and even served on the scientific board of ENICHEM.

I have always felt a special relationship with my students and postdoctoral fellows. Our joint research effort was only part of this relationship, which did not end with their leaving my laboratory. We usually keep in contact, like distant family members do, and see each other on occasion. I am proud of my "scientific family," and I hope they also have fond memories of their stay in the Olah group. My office door always was and still is open without any formality to any member of my group to discuss chemistry or any personal question or problems they want. Because Judy was an active, key member of our group during much of my academic career, my students got used to unburdening themselves more to the "kinder, gentler" Olah. They were perhaps somewhat awed by my height (6'5" or, by now, 6'4"), not always realizing that underneath the somewhat towering appearance in fact there always was a rather caring friend. Although I jealously protected my time to concentrate primarily on my research and scholarly work, I always had time for my students and colleagues. I feel this is a natural responsibility of professors and research advisors. After my Nobel prize new obligations and pressures were starting to somewhat limit my

Judy, in Munich with the Huisgens

time, and this led me to decide, as mentioned, not to take on any more new graduate students, so as not to short-change them. My research, however, continues unabated with my postdoctoral fellows. I also maintain close contact with the graduate students in our Institute, particularly those on whose guidance committees I serve, and those of the research group of my colleague and friend Surya Prakash, with whom we are continuing joint research projects.

I was blessed with an extremely fine group of graduate students and postdoctoral fellows who over the years joined the Olah group and became my wider scientific family. I have already mentioned many of those of my Cleveland years. In Los Angeles, my graduate students included Douglas Adamson, Mesfin Alemayehu, Robert Aniszfeld, Massoud Arvanaghi, Joseph Bausch, Donald Bellew, Arthur Berrier, Arwed Burrichter, Daniel Donovan, Robert Ellis, Omar Farooq, Morteza Farnia, Jeff Felberg, Alex Fung, Armando Garcia-Luna, Judith Handley, Michael Heagy, Nickolas Hartz, Ludger Heiliger, Altaf Husain, Wai Man Ip, Pradeep Iyer, Richard Karpeles, Chang-Soo Lee, Eric Marinez, Lena Ohannesian, Alex Oxyzoglou, Golam Rasul, Mark Sassaman, Tatyana Shamma, Joseph Shih, Maurice Stephenson, Qi Wang, Mike Watkins, and Chentao York.

Postdoctoral fellows of my group in Los Angeles were, among others, George Adler, Alessandro Bagno, Mario Barzaghi, Patrice Batamack, Jeno Bodis, Jorg-Stephan Brunck, Herwig Buchholz, Imre Bucsi, Jozef Bukala, Steven Butala, Young-Gab Chun, Françoise and Gilbert Clouet, Denis Deffieux, Hans Doggweiler, Ed Edelson, Ronald Edler, Thomas Ernst, Markus Etzkorn, Wolf-Dieter Fessner, Les Field, Dietmar Forstmeyer, Stephan Frohlich, Xiangkai Fu, Helmut George, Mary Jo Grdina, Paul Mwashimba Guyo, Dong-Soo Ha, Mahommed Hachoumy, Sawako Hamanaka, Toshihiko Hashimoto, Nicholas Head, Thomas Heiner, Rainer Herges, Hideaki Horimoto, Ron Hunadi, JePil Hwang, Shin-ichi Inaba, Jens Joschek, Shigenori Kashimura, Takashi Keumi, Schahab Keyaniyan, Douglas Klumpp, Norbert Krass, Ramesh Krishnamurti, Manfred Kroll, Khosrow Laali, Koop Lammertsma, Roland Knieler, Kitu Kotian, V. V. Krishnamurthy, Christian Lambert, Thomas Laube, Jean-Claude Lecoq, Stefan Lehnhoff, Xing-ya Li, Qimu Liao, Thomas Mathew, Asho Mehrotra, Alfred Mertens, Gregory

Mloston, Tadashi Nakajima, Subhash Narang, Gebhard Neyer, Peter Morton, Koji Nakayama, Alex Orlinkov, Rhuichiro Ohnishi, Don Paquin, Maria Teresa Ramos, Pichika Ramaiah, Marc Piteau, Bhushan Chandra Rao, Tarik Rawdah, V. Prakash Reddy, Christophe Rochin, Roger Roussel, Rolf Rupp, Graham Sandford, Sabine Schwaiger, Stephan Schweizer, Jurgen Simon, Brij Singh, Gunther Stolp, Derrick Tabor, Bela Török, Nirupam Trivedi, Akihiko Tsuge, Yashwant Vankar, Julian Vaughan, Qunjie Wang, Klaus Weber, Thomas Weber, Jurgen Wiedemann, John Wilkinson, An-hsiang Wu, Takehiko Yamato, Elazar Zadoc, and Miljenko Zuanic.

Over all, in my career in America, I had some 60 graduate students and 180 postdoctoral fellows. The fact that I also worked for eight years in industry and that there was a need in both Cleveland and Los Angeles to build up my research from modest beginnings probably explains the higher ratio of postdoctoral fellows to graduate students. However, because, postdoctoral fellows generally stayed only for 1–2 years, but graduate students for 3–4 years, the composition of my group was usually balanced between them. When I celebrated my 70th birthday in 1997, we had a nice reunion connected with a Symposium, which many of the former members of the Olah group attended and which was a great personal pleasure for me.

Being involved in teaching and directing research work for half a century, I always felt that mentoring my younger colleagues was an essential part of my responsibilities. In a way, one of the best judgments of any professor is how his students or associates feel about him and also how they fare in their own careers. It is not unlike how parents feel about their children. To observe with satisfaction and pride the achievements of your scientific family, which in a way is also the continuation of your own work, is most rewarding. Because I was never associated with one of the leading universities, having also spent some of my career in industry, my graduate students usually came from more modest schools than those of their peers who have gone to our premier colleges and universities. This is quite understandable and based on sound reasons. Going to Harvard, Stanford, MIT, or Caltech to earn a graduate degree in itself usually guarantees a good chance for a successful career. On the other hand, motivation and desire also count significantly. Over the years, I have seen with great satisfaction many of my students blossom and become accomplished and productive scientists, even surprising themselves with how far they were able to go. To achieve this, however, there can be no substitute for hard and dedicated work and continued learning while "doing chemistry." A research group is indeed a suitable incubator, providing the conditions and challenges for continued development. My role was that of a catalyst and facilitator, who tried to motivate but at the same time keep our joint effort oriented and on target.

I never felt it necessary to put pressure on my students to achieve goals. Drive must come from within and, of course, in the competitive environment of a research group, from friendly competition with your peers. The professor should be careful even to minimize any pressure, because in research we never really know what idea will work out or fail. Because the professor generally gets the lion's share of credit for the research achievements of his students or postdoctorals, I believe that he/she is also responsible for failures, mistakes, or any other problems arising in his laboratory. In research we all must live by high ethical and scientific standards, not only because this is the only right way but also because observations and data cannot be fudged or faked. If your work in chemistry is of any interest, others will sooner or later reproduce it, providing a way to check your results. If your experimental facts are correct you are, of course, allowed to interpret them your way. Differing interpretations can sometimes result in heated debates or controversies, but in science this eventually is resolved when new facts become known.

There is a fundamental difference between such scientific controversies and what simply can be called "scientific fraud," i.e., deliberate falsification or fudging of data. Sloppy experimental work or data keeping can also lead to questionable or incorrect conclusions, and, although these violate established scientific standards and must be corrected (as they will), they do not necessarily represent deliberate fraud. In all this, the professor has a strict personal responsibility. As he/she is getting most of the recognition for the accomplishment of the research, it is only natural that he/she must also shoulder the responsibility for any mistakes, errors, or even falsifications. It is not acceptable, for example, to say, as is done on occasion, if in a research group some overeager or misguided member falsified data for nonexistent results that were then published, that the unsuspecting professor was only tricked into what later turned out to be embarrassing false claims. If somebody in your research group comes up with a potentially significant new finding, you must make sure before anything else that the data are correct. How many times have we all experienced short-lived expectations, which were, however, soon shown to have some rather trivial reason or explanation. Concerning fraud, there is little that does not get noticed by one's immediate lab colleagues. A professor, I believe, must have always good contact with his group and be sensitive to pick up "early warning signs" if there are any reasons for concern, long before publication. In the case of results that may indicate a breakthrough of potential significance, care is, of course, even more important. In this case it is customary to repeat the work several times over, generally also by other members of the group. Checking data and avoiding any cheating or fudging must be self-controlled in the laboratory. Again, if you don't catch the mistake yourself, others will do it for you. Blaming a student or associate publicly for whatever mistakes may have been made seems equally wrong to me. This can be an internal "family matter" but is no excuse. As a parent you are responsible for your children living in your house, and you had better know what they are doing. The same is true for students in your laboratory.

In the nonclassical ion controversy discussed in Chapter 9, there was never any question on either side of the debate about the validity of the observed data, only about their interpretation. Had any of the experimental data been questioned or found to be incorrect, this would have been soon found out because so many people repeated and rechecked the data. This is the strength of science (in contrast to politics, economics, etc.), i.e., that we deal with reproducible experimental observation and data. Nevertheless, interpretation can still result in heated discussions or controversies, but science eventually will sort these out based on new results and data.

The final evaluation and judgment of any research and its significance comes from the wider scientific community upon its publication. In my research, I have always believed that there is a fundamental obligation to publish our work in leading peer-reviewed journals. This allows not only communication of the work to the wider scientific community but also its evaluation and, if needed, criticism by others around the world. No research is complete without publication of its results and conclusions. I also felt a strong obligation toward my collaborators and students, who really made the research possible and contributed fundamentally to it. Even if after a time your own drive to see your name on publications (which as in case of any other intellectual or artistic activities certainly is a factor) may be diminishing, your younger colleagues, who spent some of their best years on the research project, deserve to see their work published to receive full credit for it. Publishing scientific work is clearly an essential part of the research process. If one is not interested or prepared enough to write up and publish the results of the work in a relatively timely manner, valid questions can be raised as to whether the research should have been undertaken at all. Some criticize researchers whose publications they believe are too numerous. I may be one of these, because I have published some 1200 research papers (excluding other printed materials such as abstracts of lectures, reviews, comments, and letters). My goal, however, was never to "pad" my publication list. I myself never paid much attention to the number of my publications, although some of my friends and colleagues occasionally did comment on it. As I said earlier, I always felt and still feel that publishing is an integral part of any research or scholarly project and should be considered in this context.

Chemistry is also a practical science with a very significant industrial base. University education in chemistry, including graduate schools, however, hardly prepares one for a career in industrial chemistry. Although chemistry faculties generally are composed of very competent and capable chemists, providing a thorough chemical education, most lack industrial experience that they could convey to their students. I was fortunate in this regard that I had spent eight years in industry and thus had a realistic first-hand view to present to my students about both academic and industrial chemistry. My industrial experience, I feel, served me well also as a teacher and a mentor for my students who were going into an industrial career.

Another way in which academic chemists keep in touch with industry is through consulting. During the years I consulted at different times for Exxon, Chevron, Cyanamid, ENICHEM (Italy), and Pechiney-Kuhlman (France, when a friend, Lucien Sobel, was a research director). I enjoyed these contacts, because they kept me aware of current industrial developments and interests. At the same time, it was also rewarding to be able to suggest to my industrial friends new approaches and directions that on occasion were useful and resulted in practical applications.

An additional but more limited aspect of my consulting over the years involved some patent cases connected with my fields of expertise. The most interesting of these involved the Ziegler-Natta polymerization of olefins. It involved the work of Max Fischer, a German industrial chemist during WWII, who polymerized ethylene with $AlCl_3$ and Al powder in the presence of $TiCl_4$ under moderate pressure and obtained polyethylene. In the extensive litigation of the Ziegler-Natta polymerization the question arose of whether Fischer's work was a Friedel-Crafts type ionic polymerization or a forerunner of the Ziegler polymerization. I found the question so fascinating that subsequently we carried out some fundamental research into the underlying chemistry of $AlCl_3$/Al, which can form aluminum dichloride and related organoaluminum compounds.

I gave up most consulting after moving to Los Angeles and starting the Loker Hydrocarbon Research Institute. To build the Institute and to assure its future became my priority, and I felt that I should not dilute my effort with much private consulting. Our Institute concentrates its efforts on fundamental research and graduate training in the broad area of hydrocarbon chemistry. We decided not to accept any contract work or support from industry that would directly involve working on the specific problems of the donors. Instead, when our chemistry results in discoveries we feel may have practical use, we patent them, and the Institute subsequently welcomes arrangements with interested industry to acquire and further develop our technology for practical processes.

Starting with my work in Hungary, through my years with Dow Chemical, and during my academic career, besides scientific publications, I applied for patents on discoveries I felt represented practical significance. Over the years I obtained some 120 patents. In the Loker Institute, patents and intellectual property generated by our work has begun to contribute some support for the research of the Institute and our University.

Universities or university-related institutes, except in such fields as computer and software technology and biotechnology, are still learning how to utilize the technical discoveries and know-how of their faculty. In our field of hydrocarbon research this is a particularly difficult challenge. No one can start an oil refining or petrochemical operation in his garage or backyard. Cooperation with industrial organizations, with their technological expertise and ability to work toward commercialization, can be mutually advantageous. It is too early to say how successful we will be in the long run, but some cooperation with Texaco, UOP (in this case involving rewarding contact with an old friend, Jules Rabo, a pioneer of zeolite catalysis), and others in areas of hydrocarbon chemistry, as well as with Caltech-JPL in the field of new fuel cell development, is promising. It is certainly a new challenge and learning experience.

My work in superacid and related synthetic chemistry developed a series of useful reagents. In the mid-1970s, I helped a former graduate student of mine, Jim Svoboda, to start a small company in Cleveland called Cationics, to make and sell these cationic reagents and superacidic systems that were not readily available to the wider chemical community. He and his wife (with Judy's initial help) made a great effort, and within a short time the small company gained some recognition and helped to make our reagents better known and available. They were eventually (and still are) distributed by some of the major fine chemical and reagents companies (Aldrich, Alpha, among others). Financially, however, Cationics was hardly viable. Jim learned the hard way that money is not necessarily made by laboriously making reagents but by selling them with a large profit margin after obtaining them from other sources (including struggling small companies). Cationics moved from Cleveland to Columbia, South Carolina, was subsequently absorbed into Max Gergel's Columbia Chemicals, and eventually faded away. None of us made any money on it, despite the significant effort expended. For me, it was a learning experience not to launch a start-up company if you are, and want to stay, a scientist and do not want to become involved in business.

In my later Cleveland years, I gained some other experience in the just-emerging area of high-technology companies by being asked to join the board of directors of a company attempting to commercialize a new type of disposable personal thermometer. It was based on the color change of a series of separate dots containing eutectic mixtures of two suitable chemicals melting 0.1°C apart and encapsulated in a small plastic strip to be placed under the tongue. The idea of a disposable chemical thermometer was quite ingenious, and other applications could be foreseen, including an indicator strip registering the maximum temperature to which packaged frozen foods were exposed during storage. The company (Bio-Medical Sciences, BMS) was well financed (by some $25 million at the time for a small start-up), and its investors included the Prudential Insurance Company, the Rockefeller Brothers Trust, and Yale University. Replacement of the mercury thermometer with other devices (including electronic ones) subsequently came about. BMS itself was bought out and its device is still marketed. In any case, when I moved to Los Angeles I gave up outside interests, because I felt it proper to concentrate on my efforts at USC.

Life presents us all with many challenges. I learned early in my life to face such challenges, including just survival in the difficult and troubled WWII years in my native Hungary. The subsequent years in a poor and much destroyed country moving from one extreme regime to another were also not easy. It is said that hardship shapes your character and strengthens your spine, but after a while you feel that perhaps you need no more of it. Thus, when in 1956 my family and small research group escaped a rather dismal situation and we started a new life in America, I was rather well prepared to cope with life's challenges. I am most grateful to my adopted country, which gave me the opportunity to restart my career in chemistry and provided me and my family a new home. When I was asked years later to write a brief statement at the end of my Who's Who biographical sketch, I wrote, "America is still offering a new home and nearly unlimited possibilities to the newcomer who is willing to work hard for it. It is also where the 'main action' in science and technology remains." I still feel the same way.

We also frequently face other challenges in life. As mentioned, some twenty years ago I went through two life-threatening health crises. I came through, and I hope to be able to continue my work for more years to come. You learn from such "close encounters" what is really important in your life. I believe I owe much to having learned from hardships and challenges, and I appreciate the love and support of my family more than I can express. I also appreciate the privilege of having been able to work with wonderful students and colleagues. I am also grateful that I was able to follow my own ways and principles, with whatever shortcomings I have. From the adversities I faced and difficulties I encountered, I learned to look only forward and to make the best out of my possibilities. I never regretted or second-guessed the pathway of my life. Life anyhow is too short to worry much, and it is better to concentrate on moving forward. I also learned to better distance myself from the many inevitable small upsets and irritations we all face, which after a while turn out to be quite unimportant. In your professional work, too, you learn to differentiate what is superficial from what is really meaningful. I also learned to politely say "no" to many invitations, involvements, etc., which, however worthwhile, inevitably would distract me from pursuing my essential goals.

Scientists or, for that matter, writers, artists, performers, and athletes, cannot deny that acknowledgment of their accomplishments is satisfying and rewarding. During a long career in science, you receive different prizes and medals to recognize some of your work. I received, for example, from the American Chemical Society in 1964 its Award in Petroleum Chemistry (which recently was endowed and renamed as the "George A. Olah Award in Hydrocarbons or Petroleum Chemistry"), in 1979 the Award for Creative Work in Organic Synthesis, and in 1989 the Roger Adams Award and just recently the Arthur C. Cope Award. I also was given the Baekeland and Tolman Awards and Morley Medal, and received other recognitions such as two Guggenheim Fellowships and a senior Humboldt Award. A pleasing recognition I received was the Cotton Medal in 1996, named after Al Cotton, a leading inorganic chemist, outspoken advocate of science and education, and a good friend. Receiving recognitions from your peers, who can judge your work best, is always particularly rewarding.

It is necessary, however, to keep recognitions in proper perspective. Receiving an award or prize (even the Nobel Prize) is based on subjective judgments. No judgment or evaluation can be perfect. Life goes on, and it is frequently more revealing to observe how people continue in their field after such recognitions. I myself differentiate recognitions I received before my Nobel Prize from those that were given me afterwards. You tend, for example, to collect honorary degrees, memberships in learned societies and academies, etc. for which you probably would not have been considered otherwise. In any case, during my career I have been elected to a number of such scientific bodies, and I am proud of it. These include the U.S. National Academy of Sciences, the British Royal Society, the Italian National Academy Lincei, the Canadian Royal Society, and the Hungarian Academy of Sciences. Election to such bodies is certainly an honor, although once elected you find out that members generally take little further role in their activities.

I also received a number of honorary degrees such as from the University of Durham, in 1988 presented by its Vice-Chancellor of the time Dame Margot Fonteyn, the famous ballet dancer and a remark-

Receiving the Cotton Medal from Al

able lady. Others came from the University of Munich (1990), University of Crete (1994), my alma mater the Technical University of Budapest (1989), the Universities of Szeged and Veszprem in Hungary (1995), the University of Southern California (1995), Case Western Reserve University (1995), University of Montpellier (1996), and New York State University (1998).

Regrettably, our life, including science and higher education, is becoming increasingly bureaucratic. Talented researchers spend much of their time in committees, boards, and panels instead of pursuing their research. I have always tried to keep centered on my work and not be sidetracked, remembering a short story by Leo Szilard, the remarkable Hungarian-born scientist, included in his book *The Voice of the Dolphins*, published in the 1950s. It tells of a scientist who became a celebrity in his city and on occasion met a very wealthy man who asked him what he should do with his fortune. The scientist suggested that he should set up a foundation supporting scientific work. The millionaire answered that he would not consider it, because he intensely disliked science. After some consideration our scientist, however, came up with a solution. You should still establish your foundation, he advised, but insist that its board as well as its numerous committees should be composed of the most productive and promising scientists. To assure this, the foundation would pay them such high honoraria that nobody would decline. With all these outstanding scientists fully involved in their bureaucratic assignments, their productivity will rapidly decrease and eventually cease. Science will thus wither, and you will achieve your goal. There is much to think about in this imagined story. In fact, too many scientists are becoming increasingly less productive by allowing themselves to be overwhelmed by many commitments outside their research. In some areas such as the computer electronics and biotechnology fields, commercial interests also tend these days to overshadow science. I myself always tried to avoid such things as much as possible and stayed focused. Others may say that avoiding committees, etc. is a very selfish attitude, but are creative artists, writers, and composers any less selfish?

Finally, I have always tried to keep a healthy sense of humor, much needed in our present time. Similarly, I have managed not to take myself too seriously, only my science, about which I am quite passionate. This is useful to emphasize, because for scientists in the long run what really counts is what they have achieved, not their personality or ego. To stay productive and maybe even creative, one must maintain curiosity and enjoy the pleasure of being able to search for new understanding and of making occasional discoveries (if they still come your way). Those who follow success and recognition succumb to overblown feeling of self-importance, frequently missing out on these. It is sometimes said that Nobelists who receive their prize later in life are lucky, because after the prize one's scientific productivity generally decreases. This indeed may be the case generally, but it is possible to avoid it. I considered myself one of the lucky ones who succeeded in avoiding this trap. I believe that the key is how well you can focus on continuing your scientific work unabated, if this is really your goal. To keep my priorities focused and to be able consistently to say *no* to the many worthwhile activities the prize inevitably presented me were the most important guidelines I learned and they serve me well.

♦ ♦ ♦ ♦ ♦

My long journey, which started in my native Budapest on the banks of the river Danube and took me to the shores of the Pacific Ocean, was not always an easy one. Human nature, however, helps to block out memories of hardship and difficulties. They fade away and you look back remembering mostly the positive aspects of your life. I followed my own principles and went my own way. It helped that I inherited a strong, perhaps on occasion stubborn, nature with a determination to follow the pathway to my goals and that I worked hard to achieve them. It was and still is a rewarding life experience I shared with Judy my whole life, including our common profession. I am glad that our sons have not followed us into the area of science. George an MBA, is the treasurer of an insurance company and Ron, a physician, is practicing internal medicine in close-by Pasadena. They both have their successful careers not burdened by any comparison with their father. I was blessed to always have had the help, support, and love of my wife and family. What else could I ever have asked for?

· Appendix ·

My Previous Books for References and Additional Reading

As mentioned in Chapter 14, it was my usual practice during my career that whenever I felt that I had substantially achieved my goals and interest in a specific field of my research, I wrote (or edited) a book or comprehensive monograph of the field. The interested reader may want to consult these for further details. They also contain extensive references to my 1200 original papers as well as to reviews and chapters. If not otherwise indicated, my books were all published by my longtime publisher, John Wiley & Sons, Inc., in New York.

Introduction to Theoretical Organic Chemistry (in German), Akad. Verlag, 1960.

Friedel-Crafts and Related Reactions, Vols. I-IV (ed), 1963–1964.

Carbonium Ions (ed. with Schleyer), Vols. I-V, 1968–1972.

Friedel-Crafts Chemistry, 1973.

Carbocations and Electrophilic Reactions, 1973.

Halonium Ions, 1975.

Superacids (with Prakash and Sommer), 1985.

Hypercarbon Chemistry (with Prakash, Williams, Field, and Wade), 1987.

Nitration (with Malhotra and Narang), VCH, 1989.

Cage Hydrocarbons (ed.), 1990.

Chemistry of Energetic Materials (ed. with Squire), Academic Press, 1991.

Electron-Deficient Boron and Carbon Clusters (ed. with Wade and Williams), 1991.

Synthetic Fluorine Chemistry (ed. with Chambers and Prakash), 1992.

Hydrocarbon Chemistry (with Molnar), 1995.

Onium Ions (with Laali, Wang, and Prakash), 1998.

Research Across Conventional Lines, Collection of papers of George Olah and Commentary (ed. with Prakash), World Scientific Publ., Singapore, 2001 (in press).

Index